AF834894

Nanoscience and Technology

Nanoscience and Technology

V. S. Muralidharan
Central Electrochemical Research Institute
Karaikudi-630 006

&

A. Subramania
Alagappa University
Karaikudi-630 003

CRC Press
Taylor & Francis Group
Boca Raton London New York

CRC Press is an imprint of the
Taylor & Francis Group, an **informa** business
A TAYLOR & FRANCIS BOOK

Nanoscience and Technology
V. S. Muralidharan and A. Subramania

First published 2009 by Ane Books Pvt. Ltd. for CRC Press

Published 2024 by CRC Press
2385 NW Executive Center Drive, Suite 320, Boca Raton FL 33431
4 Park Square, Milton Park, Abingdon, Oxon, OX14 4RN

CRC Press is an imprint of Taylor & Francis Group, LLC

© 2009 Taylor & Francis Group, LLC

Reasonable efforts have been made to publish reliable data and information, but the author and publisher cannot assume responsibility for the validity of all materials or the consequences of their use. The authors and publishers have attempted to trace the copyright holders of all material reproduced in this publication and apologize to copyright holders if permission to publish in this form has not been obtained. If any copyright material has not been acknowledged please write and let us know so we may rectify in any future reprint.

Except as permitted under U.S. Copyright Law, no part of this book may be reprinted, reproduced, transmitted, or utilized in any form by any electronic, mechanical, or other means, now known or hereafter invented, including photocopying, microfilming, and recording, or in any information storage or retrieval system, without written permission from the publishers.

For permission to photocopy or use material electronically from this work, access www.copyright.com or contact the Copyright Clearance Center, Inc. (CCC), 222 Rosewood Drive, Danvers, MA 01923, 978-750-8400. For works that are not available on CCC please contact mpkbookspermissions@tandf.co.uk

Trademark notice: Product or corporate names may be trademarks or registered trademarks and are used only for identification and explanation without intent to infringe.

British Library Cataloguing in Publication Data
A catalogue record for this book is available from the British Library

ISBN: 978-1-4200-8962-2 (hbk)

Visit the Taylor & Francis Web site at
http://www.taylorandfrancis.com

and the CRC Press Web site at
http://www.crcpress.com

For distribution in rest of the world other than the Indian sub-continent

Dedicated to my Teachers
– A. Subramania

Dedicated to my grandson, Ajay Ramkarthik
– V. S. Muralidharan

Preface

Sir Richard Feynman stimulated the world in 1959 to pay attention to delve into the question of fabricating materials and devices at the atomic/molecular scale. Only in 1980's scientists of IBM moved 35 Xenon atoms to spell out 3 letter IBM logo. Since then progress during the two decades in nanoscience and technology is catastrophically high.

An elementary approach to a highly complicated emerging field is the main concern of this book. Numerous research publications over the past two decades on Nanoscience and Technology added to the profound knowledge. In popular articles on nanoscience and technology many serious readers would have come across an involved mosaic of theories taken from quantum physics and chemistry. To facilitate them to understand an introductory chapter on foundations in nanoscience is given. Difficulties in writing a book on the emerging field arises from the large number of publications which handled various approaches in synthesis, materials characterization and no doubt terminologies. An earnest attempt is made to remove jargons not sacrificing the intrinsic features of the smail world."There is plenty of space at the bottom".

The growth of nanoscience and technology was due to the untiring research by numerous scientists around the globe. An elementary book like this would be no measure to be little their work. The book is only a "start"; marathons are yet to follow. There are numerous ways to present this field and we present the subject in four parts.

First part introduces the terminologies in Nanoscience. Quantum structures, nanoclusters, Carbon nanotubes inorganic nanotubes, nanowires, organic nanocrystals and nanofibers are presented. Various methods for the synthesis of metal colloids, nanoclusters, polymer supported clusters, nanotubes, carbon nanotubes, nanowires, nanorods, nanocrystalline materials, oxide nanoparticles, nanointermetallics, nanocomposites and polymeric nanofibers are presented in part two. As the engineering of nanoparticles emerges, there is an important concern to characterize them and to standardize them. Characterized nanoparticles have to be standardized. Third part presents characterization and standardization. Structural elucidation methods viz., X-ray diffraction and absorption, particle size determination, structure of surfaces, spectroscopy, luminescence and microscopy are given in detail. Standards of nanometerology are also presented at the end. Last part of this book presents fewer applications of nanoscience, nanobiology, nanocatalysis, nanoelectrodes, nanomachines, nanoswitches, nanocomputers and nanofilters.

Web searches were made using various search engines; visiting them added complexity rather than clarity. Drowned in a sea of information we really starved of real fruitful knowledge to offer. Some useful sources thus obtained is presented as bibliography. The reader has to have basic knowledge of atomic structure, quantum chemistry and semiconductor physics. The book is most suitable for undergraduate courses in US and Canadian universities and for post graduate courses in Indian and SAARC countries.

— Authors

Acknowledgement

At the outset let me offer my sincere prayers to God for the successful completion of this joint endeavor. I am thankful to my first daughter, Smt. Sowmiya Ramkarthik and Sri Ramkarthik who supported and inspired me to write this book during my stay at Folsom, CA, USA. My second daughter Shambavi, offered me suggestions and encouragement periodically. My grandson, Ajay deserves special appreciation who kept me awake in early mornings; his smiles and laughter helped me to overcome tiredness. Writing a book and raising a middle class Indian family have many things in common; both respond to recurring, subliminal desire and legacy to leave behind.

My thanks are due to the authorities of Central Electrochemical research institute, Karaikudi for permitting me to stay in USA for a year. I am indebted to M/s Ane Books Pvt. Ltd. and the readers for selecting this book. My success is always due to my wife Smt. Sasikala who is with me during various stages of fortune and misfortunes for more than 3 decades.

— V. S. Muralidharan

Acknowledgement

First of all, I express my deep sense of gratitude to **God almighty** for the fruition of this joint venture successfully. I wish to express my sincere thanks to Dr. C.N.R. Rao, National Research Professor, Jawaharlal Nehru Centre for Advanced Scientific Research, Bangalore – 560 064, who inspired me to do research on Advanced Materials Science and Technology. My sincere thanks to Professor. Dr. T. Vasudevan, Ex-Head, Department of Industrial Chemistry, Alagappa University, Karaikudi and Dr. P. Manisankar, Professor & Head, Department of Industrial Chemistry, Alagappa University, Karaikudi for their constant support and encouragement. I express my sincere thanks to Dr. P. Kanniappan, Former Vice-Chancellor and Dr. P. Ramasamy, Present Vice-chancellor and Dr. R. Dhandapani, Registrar, Alagappa University, Karaikudi for their kind support and permission to publish this book.

My special thanks to my wife Mrs. S. Raje Suganeya who has assisted and supported me in many ways during the preparation of this book. Without her involvement, the book would not have reached readers. Last but not least, I express my profound thanks to M/s Ane Books Pvt. Ltd. for bringing out this book in an excellent way.

— A. Subramania

Contents

CHAPTER 1

Foundations in Nanoscience

1.1 NANOSCIENCE AND TECHNOLOGY

The science of nanomaterials has created great excitement and expectations in the last decade. Now "Nano" has moved from the world of the future to the world of the present. The prefix 'nano' means one billionth of a meter. One nanometer (nm) is 10^{-9} m. Nanoscience is the study of fundamental principles of molecules and structures with at least one dimension roughly is 1 to 100 nm. These structures are known as nanostructures. Nanotechnology is a creation and exploitation of materials with structural features in between those of atoms and bulk materials. Anything smaller than a nm in size is just a loose atom or small molecule floating in space as a little speck of vapour. So nanostructures are not just smaller than anything we have made before. At the nanoscale, fundamental properties change. For example, a nanoscale wire or circuit component does not necessarily obey Ohm's law. When we reach nanoscale, everything will change including the gold's colour, melting point and chemical properties. The reason for this change has to do with the nature of the interactions among the atoms that make up the gold. Interactions that are averaged out of existence in the bulk material. Nanogold does not act like bulk gold.

Nanofabrication involves two ways. Top-down fabrication is to start with a chunk of the material like gold and successive cutting it to nanoscale. Conversely bottom-up fabrication is to start with individual atoms and building up to a nanostructure. The tiny gold nanostructures are called quantum dots or nanodots, because they are roughly dot shaped and not have dimensions at nanoscale. Nanoscale gold particles are known in medieval and Victorian churches. Nanoscale gold particles can be orange, purple, red or greenish depending on their size.

Moore's Laws

The trends in the development of computer hardware have been remarkably steady for the last 50 years. Plotted on semilog paper as a function of year such parameters as the number of atoms required to store one bit, the number of dopant atoms in a transistor, the energy dissipated by a single logic operation, the resolution of the finest machining technology and others, have all declined with remarkable regularity. From relays to Vacuum tubes to transistors to integrated circuits to very large scale integrated circuits (VLSI), we have seen steady declines in the size and cost of logic elements and steady increases in their performance. If extrapolate these trends we find, they reach interesting values in the year 2010 to 2020 time frame. The number of atoms required to store one bit in a mass memory device reaches one. The number of dopant atoms in a transistor reaches one. The energy dissipated by a single logic operation reaches kT for T=300K. This is roughly the energy of a single molecule bounding around at room temperature. The finest machining technology reaches a resolution of roughly an atomic diameter. To quote Alan Kay "the best way to predict the future is to invent it".

Gordon E. Moore, one of the founders of Intel Corporation described in the article titled "cramming more components onto integrated circuits". Electronics, April 19, 1965, pages 114-117, described two laws :

1. First law states that the amount of space required to install a transistor on a chip shrinks by roughly half every 18 months. This means that the spot that could hold one transistor 15 years ago, can hold 1000 transistors today.
2. Second law states that the cost of fabricating a chip doubles with every other chip generation or roughly every 36 months.

Moore also stated that we are beginning to approaching some real limits. It gets difficult to make smaller and smaller structures. "We just exhaust the capabilities of optical systems; we need something different to make them smaller yet". A side from nanoscale electronics, one part of which due to its focus on molecules is often called molecular electronics.

Nanotechnology

At the molecular scale, the idea of holding and positioning molecules is new and shocking. However as early as 1959, Richard Feynman, the Nobel Laurete, said that nothing in the laws of physics prevented us from arranging atoms the way we want... it is something in principle that can be done; but in practice it has not been done because we are too big. He delivered a talk, "there is plenty of room at the bottom", in 1959.

Electron was discovered early in the 20th century. It is very light and has a negative charge. The simplest picture of an atom consists of a dense heavy nucleus with a positive charge surrounded by a group of electrons that orbit the nucleus and that (like all electrons) have negative charges. All atoms are roughly 0.1nm in size. Helium is the smallest naturally occurring atom is 0.1nm and Uranium is the order 0.22nm. All atoms are smaller than the nanoscale.

Self assembly is the art and science of arranging conditions so that the parts themselves spontaneously assemble into the desired structure. It is the path to nanotechnology. Molecular mechanics allows computational modeling of the positions and trajectories of the nuclei of individual atoms without an undue computational load. The ability of one molecule to attract and bind to another is often referred to as molecular recognition. Nanotechnology involves building from the bottom and so molecular recognition is important. Atoms on a small scale behave like nothing on a large scale for they satisfy the Laws of quantum mechanics. They obey different laws. At the atomic level, new kinds of forces and possibilities operate. To fix atoms and to arrange them, one needs tools to see, measure and manipulate at the nanoscale.

Some of the first tools to help launch the nanoscience revolution were the so called scanning probe instruments. All types of scanning probe instruments are based on an idea developed at the IBM Laboratory at Zurich. In scanning probe measurements, the probe (also called a tip) slides along a surface in the same way one's finger does on a wood or silk. In scanning tunneling microscopy (STM) the amount of electrical current flowing between a scanning tip and a surface is measured. Depending on the way the measurement is done, STM can be used either to test the local geometry (how much the surface protrudes locally) or to measure the local electrical conducting characteristics. STM was actually the first of the scanning probe methods to be developed and Gerd Binning and Heinrich Rohrer shared Nobel Prize for this. In magnetic field microscopy (MFM), the tip that scans across the surface is magnetic. The MFM tip works in a similar way to the reading head on a hard disk drive. Scanning probes were used to manipulate the individual molecule beads on the molecular structure. Small molecules can be moved on a surface either by pushing on them or by picking them up off the surface onto a scanning tip that moves around and puts them back down. Scanning probe surface assembly is inherently very elegant, but it suffers from cost and time.

In 1959 Feynman said, "I am not inventing anti gravity which is possible someday only of the laws are not what we think. I am telling you what could be done if the laws are what we think, we are not doing it simply because we have not gotten around it". Tinniness is incredible. In 1990 Donald Eigler of IBM Alanadan and E.K. Schweizer from Fritz-Herber have first moved 35 Xenon atoms to spell out 3 letter IBM logo a top a crystal of nickel (**Figure 1.1**). As Feynman said, "we have been content to dig in the ground to find materials. We heat them and do things on a large scale with them and we hope to pure substance with just so much impurity

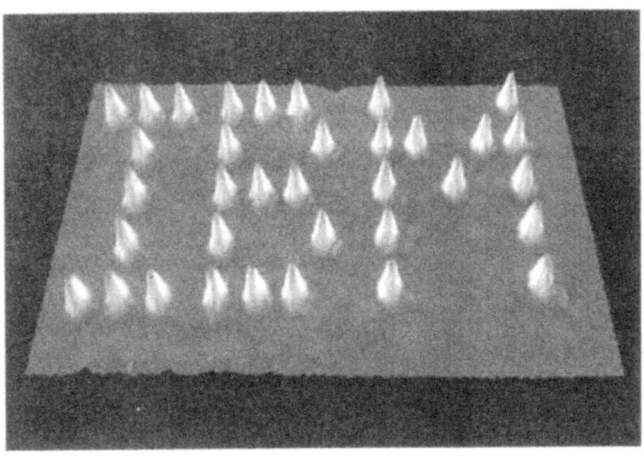

Fig. 1.1 Positioning single atoms with a scanning tunneling microscope.

and so on. But we must always accept some atomic arrangement that nature gives us. I can hardly doubt that when we have some control of the arrangement of things on a small scale we will get an enormously greater range of possible properties that substances can have and of different things we can do."

A lithograph is an image that is produced by carving a pattern on the stone, inking the stone and then pushing the inked stone. Nanoscale lithography really cannot use visible light because the wavelengths of visible light is at least 400nm. To write, atomic force microscope tips are used as nanopens. Dip-pen nanolithography (DPN) is named after 19th century dip-pen used in school rooms. DPN was developed by Chad Mirkin and co-workers at Northwestern University. In DPN, a reservoir of "ink" (atoms or molecules) is stored on the top of the scanning probe tip which is manipulated across the surface, leaving lines and patterns behind. E. beam lithography can be used to make structures at the nanoscale. Rick Van Duyne's group used a technique known as Nanosphere Liftoff Lithography. It has nice features : many sorts of surfaces and metals and molecules can be used. Nanotechnology gives us tools to play with ultimate toy box of nature atoms or molecules. Dip-pen (similar to dip-pen of ancient times) nanolithography enables to write letters of 400 nm (**Figure 1.2**).

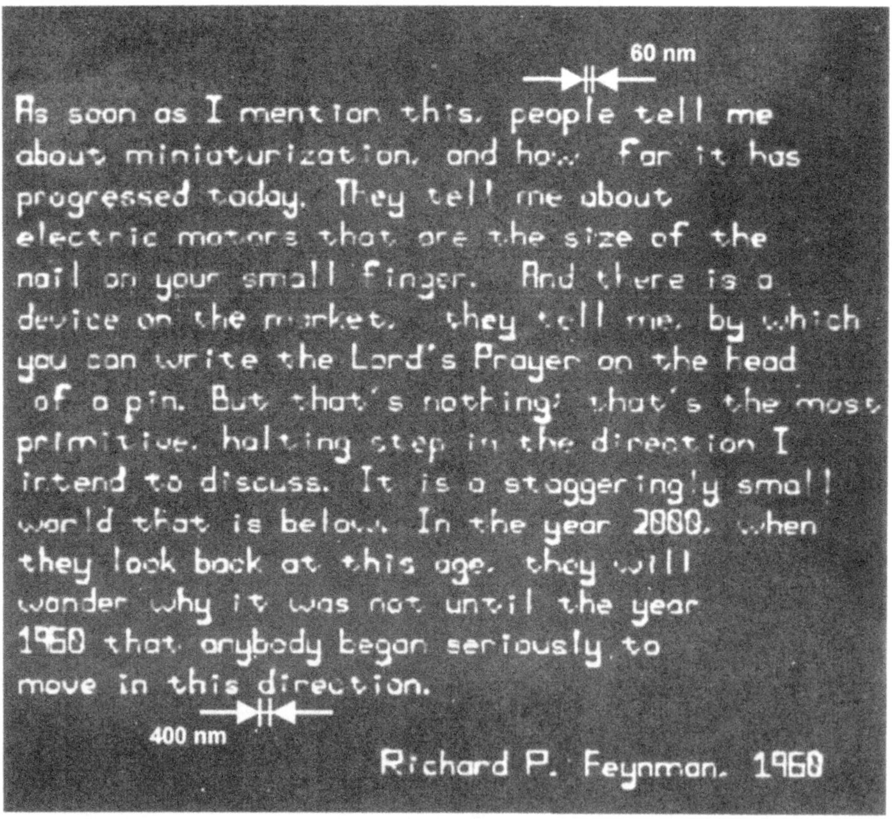

Fig. 1.2 Dip-pen nanolithography.

Self assembly is a nanofabrication technique. The forces involved in self assembly are generally weaker than the bonding forces that hold molecules together. They correspond to weaker aspects of coulombic interactions. In self assembly, the nanobuilder introduces particular atoms or molecules onto a surface or onto to a preconstructed nanostructure. The molecules then aliegn themselves into particular positions, sometimes forming weak bonds and forming strong covalent ones in order to minimize the total energy. Self assembly is almost certainly going to be the preferred method for making large nanostructure arrays such as the computer memories and computer logic that must be prepared, if Moore's law is going to continue to hold true beyond the next decade. Sam stupp at North Western University USA has two levels of self assembly. First is the self assembly of long complex molecules called "rod coils" products (the mushroom like nanostructure). Then the nanostructure themselves self assemble to produce a surface coating that makes the glass slide either hydrophilic or hydrophobic. Crystal growth is another sort of self assembly "Silicon boules," the blocks used for making microchips, are made or 'drawn' in this way. Charles Lieber and his group at Harvard University used nanoscale crystals to seed long wire like single crystals of carbon nanotubes as well as compounds such as InP or GaAs.

In modern biotechnology, a combination of specific short DNA sequences and self assembly is used, in which a single DNA strand binds to another single DNA strand. Nanostructures must be assembled. Two of the popular nanostructures are so called 'Carbon nanorods", prepared first by Sumino Iijima in Tokyo and nanorods made out of silicon. Non-carbon tubes are usually called Nanowires. Engineers in Phaedon Avouris's group at IBM used nanotubes to craft usable transistors.

Understanding how to construct nanoscale materials and to study their properties are central to nanoscience. The use of quantum mechanical theory to predict actual molecular structures is one of the triumphs of this century.

Future of Nanotechnology

For a copper wire, resistivity decreases linearly as temperature goes down. The resistivity of cryogenic wire (aluminium) exhibits a step change at a specific low temperature but never becomes zero at 0K. The resistivity of superconductor wire decreases as temperature goes down and abruptly becomes zero at a certain low temperature, T_e, called the Curie temperature. When a certain physical parameter as a function of size, for ordinary materials the parameter may change linearly as size becomes small. With some exordinary materials the parameter may exhibit a step change at a certain small size as in the cryogenic wire. This is called "Small size effect".

Compressed ferrous alloy powder developed by Sumitomo electric in 2001 shaped in coin exhibits high electromagnetic wave absorption in the microwave frequency region by resonance. They are used in cell phones, game consoles, BS/CS converters and PCs.

Diamond is not a substantially self structured nanomaterials unlike fullerene. Diamond has a rigid structure that gives it hardness, very high thermal conductivity and high acoustic velocity. It is a semi-conductor and used in optical, semi-conductor and electron emission devices. It is a future nanomaterial and 2nm size has about 10 atoms of carbon. This is nearly equal to the radius of a carbon nanotube, 2nm size, is achieved by top-down nanotechnology. Electron emission from the tips is increased by almost 1million times at the applied electric field of 1.0V/μm as compared with that from the flat diamond surface. This is another example of nanosize effect.

The size of triode vacuum tube is of the order of several cm. Diamond nanoemitter has the reasonable electron emission at 30°C. Vacuum microelectronic device (VMD) has the features where the electrons travel in vacuum space, the mobility and the velocity of electrons are much larger then any other semi conductor materials.

1.2 QUANTUM STRUCTURES

When the size of a material reduced from centimeters to 100nm, there is a sudden change in properties. If one dimension is reduced to the nanorange while the other two dimensions can remain large, then the structure is known as a quantum well. If two dimensions are so reduced and one remains large, the resulting structure is Quantum wire. When all the three dimensions are in nm range then the structure is a quantum dot. There are two major approaches of preparation. One is the bottom-up approach is to collect, consolidate and fashion individual atoms and molecules into the structure. This is carried out by a series of chemical reactions. Top-down method starts with a large scale object or pattern and gradually reduces its dimensions. This can be accomplished by a technique known as "Lithography".

The first step of the lithographic procedure is to place a radiation-sensitive resist on the surface of the sample substrate. The sample is then irradiated by an electron beam in the region where the nanostructure pattern or a scanning electron beam that strikes the surface only in the desired region. The radiation chemically modifies the exposed area of the resist so that it becomes soluble in a developer. The third step in the process is the application of the developer to remove the irradiated portions of the resist. The fourth step is the insertion of an etching mask into the hole in the resist and the fifth step consists of in lifting off the remaining parts of the resist. In the 6th step, the areas of the quantum well not covered by the etching mask are chemically etched away to produce the quantum structure. Finally the etching mask is removed to provide the desired quantum structure.

Lithography employs various sources of radiation. Electron beam, neutral atom beams, charged ion beams (e.g Ga^+) or electromagnetic radiation such as visible light, UV, or X-rays. When laser beams are used frequency doublers and quadroplers can bring the wavelength into a range (λ~150nm) that is convenient for quantum dot fabrication. Photochemical etching can be applied to a surface activated by laser light.

The lithographic technique can be used to make more complex quantum structures than the quantum wire and dot. One can produce the 24 quantum dot array consisting of 6 columns each containing 4 stacked quantum dot arrays. It has been found that the arrays produce a greatly enhanced photoluminescent output of light.

Size and Dimension Factors

GaAs is a typical material for comparison. The unit cell has four Ga and four As atoms. The lattice constant is a = 0.565nm and the volume of the unit cell is 0.180nm^3. Each Ga atom is in the centre of a tetrahedron of As atoms corresponding to the grouping $GaAs_4$ and each arsenic atom has a corresponding configuration $AsGa_4$. There are about 22 of each atom type per cubic nanometer and a cube shaped quantum dot 10nm on a side contains 5.56×10^3 unit cell. GaAs is FCC and if the initial cube is taken the nanostructures containing n^3 of these unit cell are built up, then it can be shown that the number of atoms N_S of the surface, the total number atoms N_T and this size are dimension d of the cube are given by $N_S = 12^{n2}$; $N_T = 8n^3 + 6n^2 + 3n$; d = na = 0.565n; here a = 0.565nm. **Table 1** summarises zinc blende type nanosemiconductor properties. The large % of atoms on the surface for small n is one of the principle factors that differentiates properties of nanostructures from those of the bulk material. FCC nanoparticle has a greater % of its atoms on the surface for the same total number of atoms particle. This is expected because only one of the two types of atoms in the GaAs structure contributes to the surface.

Full-shell Clusters		Total Number of Atoms	Surface Atoms (%)
1. Shells	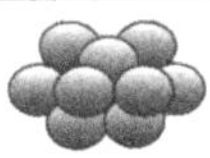	13	92
2. Shells		55	76
3. Shells		147	63
4. Shells	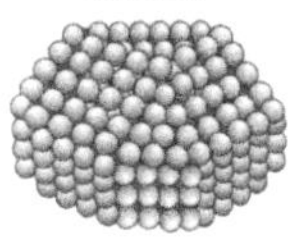	309	52
5. Shells	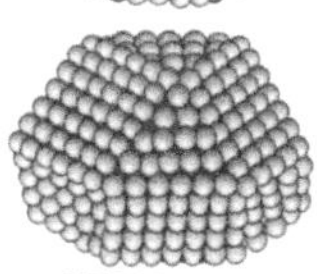	561	45
6. Shells	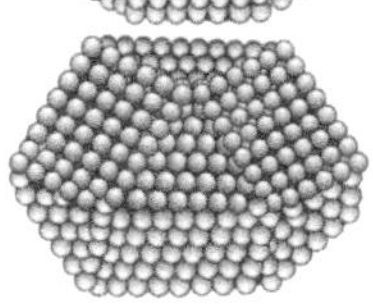	1415	35

Fig. 1.3 Relationships between shells, total number of atoms and surface atoms in %.

For clusters, total numbers of atoms increased from 13 to 1415 as one increases shell numbers from 1 to 7. However the % of surface atoms decreases with increase in shell size (**Figure 1.3**). Spherical iron nanoparticles exhibit interesting properties. The percentage of bulk atoms decreases, % surface atoms increases when the nano crystals size reduces to less than 5nm (**Figure 1.4**). The level of doping of a semiconductor and size effect have been related. For typical doping levels of 10^{14} to 10^{18} donors / cc a quantum dot cube 100nm on a side would have on the average from 10^{-1} to 10^{3} conduction electrons. The former value of 10^{-1} electrons / cc means that on the average only one quantum dot in 10 will have one of these electrons. A smaller quantum dot cube only 10nm on a side would have on the average one electron for the 10^{18} doping level and be very unlikely to have any conduction electrons for the 10^{14} doping level.

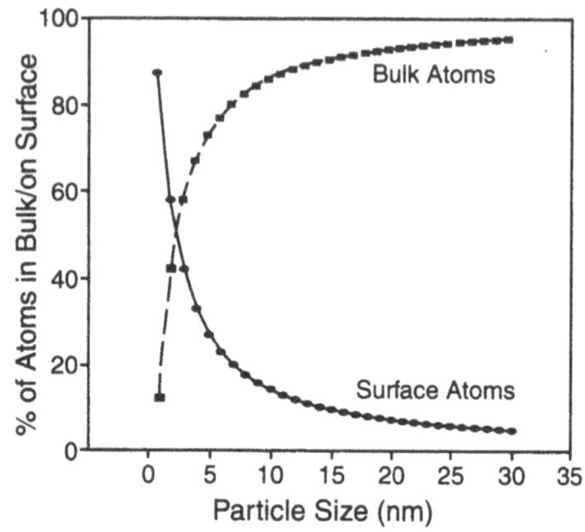

Fig. 1.4 Spherical iron nanocrystals—particle size vs % atoms in bulk/on surface.

Conduction electrons are delocalised. All the wire dimensions are very large compared to the distances between atoms. As size reduces the delocalization is impeded and electrons are confined. A copper plate of 10cm wide, 10cm long and 3.6nm thick, has only 10 unit cells. This means that 20% of the atoms are in unit cells at the surface. The conduction electrons would be delocalized in the plane of the plate but confined in the narrow dimension, a configuration referred to as a "Quantum well". A quantum wire is a structure that is long in one dimension but its diameter is in nm. The electrons are delocalized and move freely along the wire but are confined in the transverse directions. Finally a quantum dot exhibits electron confinement in all directions.

For various quantum structures (zero, 1D, 2D, and 3D structures) density of states variation with energy is shown in **Figure 1.5a**. For these structures wave function models are shown in **Figure 1.5b.**

Table 2 summarizes the conduction electron content and electron confinement.

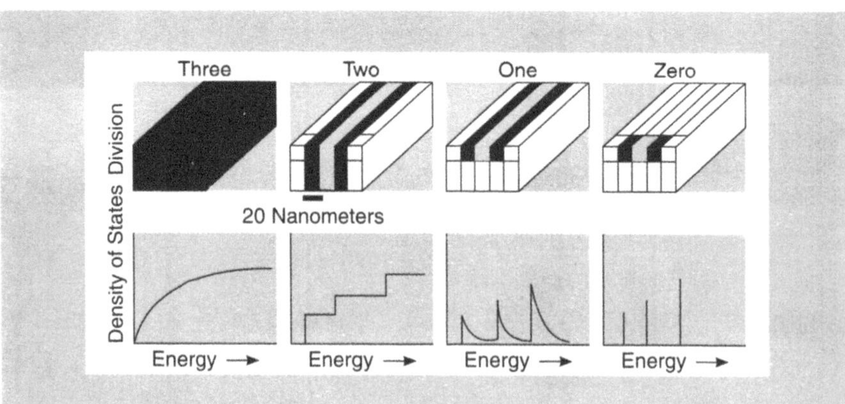

Fig. 1.5(a) Quantum structures—density of states vs energy.

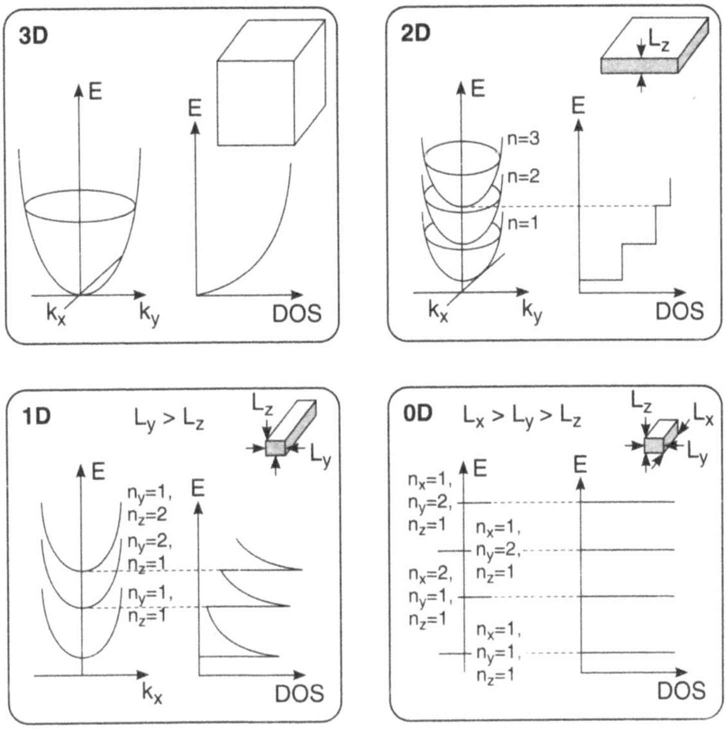

Fig. 1.5(b) Quantum structures—wave function models.

TABLE 1. Number of atoms on the surface N_S, number in the volume N_V and % of atoms N_S/N_V on the surface.

n	Size (nm)	Total number of atoms	Number of surface atoms	% of atoms on the surface
2	1.13	94	48	51.1
3	1.7	279	108	38.7
4	2.26	620	192	31.0
5	2.83	1165	300	25.8
6	3.39	1962	432	22.0
10	5.65	8630	1200	13.9
15	8.48	2.84×10^4	2700	9.5
25	14.1	1.29×10^5	7500	5.8
50	28.3	1.02×10^6	3×10^4	2.9
100	56.5	8.06×10^6	1.2×10^5	1.5

TABLE 2. Relationship between size and electron contents.

Quantum	Size structure	Electron content	Confinement direction
Bulk	—	$10^{14} - 10^{18}$ cm^{-3}	0
Quantum well	10 nm	$1 - 10^4$ μm^{-2}	1 (z)
	100 nm	$10 - 10^5$ μm^{-2}	
Quantum wire	10 × 10 nm cross section	$10^{-2} - 10^2$ μm^{-1}	2 (x, y)
Quantum dot	10 nm on a	$10^{-4} - 1$	3 (x, y, z)
	side 100 nm	$10^{-1} - 10^3$	

Fermi Electrons

The electrons in a metal act as a gas called a Fermi gas. In a quantum mechanical description the component of the electron's momentum along the x direction P_X has the value $P_X = \hbar k_X$, where $\hbar = \frac{h}{2\pi}$ (h is the Planck's constant). Each electron has K_X, K_Y and K_Z values which form a lattice in K space known as reciprocal space. At absolute zero temperature, the electrons of the Fermi gas occupy all the lattice points in reciprocal space out to a distance K_F from the origin K = 0 corresponding to a value of the energy called Fermi energy E_F. The number of conduction electrons with a particular energy depends on the value of the energy and also on the dimensionality of the space. This is because in one dimension the size of Fermi region containing electrons has the length $2K_F$ in two dimensions it has the area of the Fermi circle πK_F^2 and in three dimensions it has the volume of the Fermi sphere $4/3\pi K_F^3$. If we divide each of these Fermi regions by the size of the corresponding K space unit cell and make use of $E_F = (h^2 K_F^2/2m)$ to eliminate K_F. Now the dependence of the number of electrons on E is got when N (E) is plotted energy, the density of states can be obtained. The density of states decreases with increasing energy for one dimension, is constant for two dimensions and increases with energy for three dimensions.

When the dimensions of conductor diminishes to a nanoregion, movement of unconfined (free) electrons decreases. The electrons become sequestered in a "potential well" which is a square with very sharp boundary. An infinitely deep square potential well of width 'a' in one dimension the coordination x has the range of values $-a/2 \le x \le a/2$ inside the well and the energies there are given by

$$E_n = [\pi\hbar^2/2ma^2]n^2 = E_o n^2$$

where $E_o = \pi\hbar^2 / 2ma^2$ is the ground state energy and n assumes 1, 2, 3,.......n. The energy of a two dimensional infinite rectangular square well. $E_n = (\pi\hbar^2/2ma^2)(n_x^2 + n_y^2) = E_o n^2$ depends on two quantum numbers n_x = 0, 1, 2, 3 .....; n_y = 0, 1, 2, 3 ...... where $n^2 = n_x^2 + n_y^2$. Some nanostructures of technological interest exhibit both potential well confinement and Fermi gas delocalization. Confinement in one or two dimensions and delocalization in two/one dimensions will be interesting.

Partial confinement of electrons occurs when size is reduced. In the three dimensional Fermi sphere the energy varies from E = 0 at the origin to $E = E_F$ at the Fermi surface and similarly for the one and two dimensional analogs. When there is confinement, the conduction electrons will distribute themselves among the corresponding potential well levels that lie below the E_F along confinement coordinate directions, in accordance with their respective degenaricies d_i and for each case the electrons will delocalize with remaining dimensions by populating Fermi gas levels in the delocalization direction of the reciprocal lattice. (**Table 3**).

Phonons or Quantized lattice vibrations also have a density of states. D_{pH} (E) that depends on dimensionality. Specific heat of a solid C varies with size. The heat is the amount that excites lattice vibrations and this depends on the phonon density of states D_pH (E). At low temperatures there is also a contribution to the specific heat C_{el} of a conductor arising from the conduction electrons. The component of the magnetic susceptibility arising from the conduction electrons is called the Pauli susceptibility, $X_{el} = \mu_B^2$ D (EF). μ_B is the unit magnetic moment called the Bohr magnetron, λ_e is characterized by its proportionality to the electronic density of states D(E). Photo emission spectroscopy gives information about density of states.

TABLE 3. Number of electrons, density of states and energy for electrons.

Material	N(E)	D(E)	Dimensions	
			Delocalised	Confined
Dot	$K_0\ \Sigma d_i \theta\ (E - E_{iw})$	$K_0\ \Sigma di\delta\ (E - E_{iw})^2$	0	3
Wire	$K_1\ \Sigma d_i\ (E - E_{iw})^{1/2}$	$1/2\ K_1\ \Sigma di\ (E - E_{iw})$	1	2
Well	$K_2\ \Sigma d\ (E - E_{iw})$	$K_2\ \Sigma di$	2	1
Bulk	$K_3\ (E)^{3/2}$	$3/2\ K_3\ (E)^{1/2}$	3	0

Excitons : When an atom at a lattice site loses an electron, the atom acquires a positive charge that is called a hole. If the hole remains localized at the lattice site and the detached negative electron remains in its neighbourhood. It will be attracted to the positively charged hole through the coulomb interaction and can become bound to form a hydrogen type atom. This is called a Mott-Wannier type of exciton. The coulomb force of attraction between two charges $Q_e = -e$ and $Q_h = +e$ separated by a distance 'r' is $F = -ke^2 / \varepsilon r^2$ where e is the electronic charge, k is a universal constant and ε is the dielectric constant of the medium. The exciton radius is taken as an index of the extent of confinement experienced by a nanoparticle. Two limiting regions of confinement that can be identified on the basis of the ratio of the dimension, d, of the nanoparticle to the exciton radius, a_{eff}, the weak confinement regime with $d>a_{eff}$ (but not $d>>a_{eff}$) and the strong confinement regime $d < a_{eff}$. The more extended limit $d>> a_{eff}$ corresponds to no confinement. Under weak confinement, the exciton can undergo unrestricted translation motion and for strong confinement this motion is restricted. With decrease in size, electron hole interaction increases. As a result the energy splitting becomes greater between the radiative and nonradiative exciton states. An optical index of this confinement is the blue shift (shift to higher energies) of the optical absorption edge and the exciton energy with decreasing nanoparticle size. At room temperature, the exciton features in the absorption spectra are seen.

Artificial Atoms

A conventional field electron transistor is a switch that turns on when electrons are added to a semiconductor and turns off when they are removed. When electrons are confined in a small volume and communicate with the electrical leads by leading, all this changes. This is a single electron transistor (SET). In 1989, Scott-Thomas fabricated first semiconductor SET in narrow Si field effect transistors. In this case the tunnel barriers were produced by interface charges. Meirav and others constructed AlGaAs. In most cases the potential confining the electrons in a SET is of sufficiently low symmetry that one is in the regime of quantum chaos: the only quantity that is quantized is the energy. There is deep analogy between confined electrons and atoms. Whereas natural atoms are studied by adding, removing or exciting electrons with light, these artificial atoms typically have such small energy scales that they are best studied by measuring the voltage and current resulting from tunneling between the artificial atom and nearby electrodes. A schematic diagram of single electron transistor is shown as **Figure 1.6**.

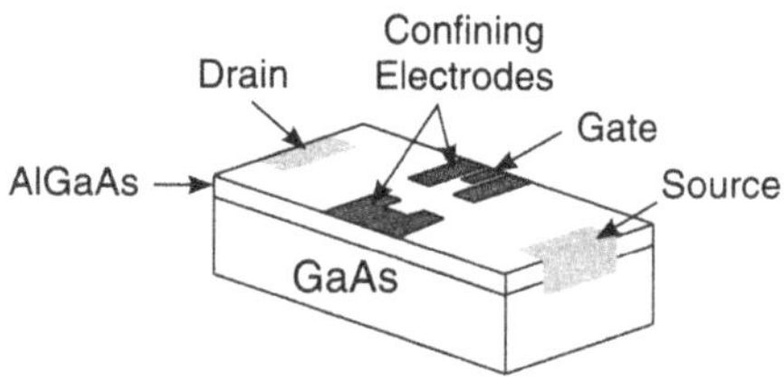

Fig. 1.6 Schematic diagram of SET.

It consists of a semiconductor, in this case GaAs separated from metal electrodes by an insulator in this case AlGaAs. The AlGaAs is doped with Si which denotes electrons. These fall into the latter material. The resulting positive charge on the Si atoms creates a potential that holds the electrons at the GaAs/AlGaAs interface creating a two dimensional electron gas (2DEG). The source and drain contacts allow one to drive electrons from an external circuit through the 2DEG. The 2DEG is confined perpendicular to the GaAs/AlGaAs interface and the confinement in the other two directions is accomplished with electric fields imposed by very small confinement electrodes. The negative voltage repels electrons from underneath the confinement electrodes and creates saddle point barriers under the constrictions. A constant voltage results a fixed confinement of potential. On an additional electrode, the gate, as the voltage is varied potential of the electrons confined in the potential well changes. These type SET's have about 50 electrons confined to a droplet of ~100nm diameter. The GaAs/AlGaAs structure is usually grown with molecular epitaxy. The electrodes are fabricated using electron beam lithography.

When the voltage on the gate electrode is increased, the potential minimum, in which the electrons are trapped becomes deeper. This causes the number of trapped electrons to increase. However unlike a conventional transistor in which the charge increases continuously, the charge in the trap increases in discrete steps. The conductance is measured by applying a very small voltage V_{ds} between drain and source, small enough that the current is proportional to V_{ds}. The conductance increases and decreases by several orders of magnitude almost periodically in V_g. A calculation of the capacitance between the gate electrons showed that the voltage between two peaks or two valleys is just that necessary to add one electron to the droplet. The name "Single Electron Transistor" comes from the observation that the transistor turns on and off again every time a single electron is added to it.

For current to flow the number of electrons on the droplets must fluctuate say between N and N + 1. Thus the Nth peak in the conductance occurs when the state of the droplet containing N electrons is in equilibrium with the state containing N + 1 electrons. Were the gate only electrode contributing to the electrostatic energy of the droplet, the gate voltage at which the Nth peak occurred multiplied by the charge of the electron "e" would be the energy difference between the two states. Since there are several electrodes near the droplet, the energy change caused by V_g is αeV_g where $\alpha = C_g /C$ is the ratio of the gate capacitance to the total capacitance. Therefore a conductance peak occurs when $\alpha_e V_g (N) = E (N+1) - E (N)$ apart from a constant, where E (N) is the total energy of the droplet with N electrons. Predicting the position of the Nth peak requires a model for the total energy of an artificial atom with N electrons.

Of the many models coulomb blockade model treats the droplet of confined electrons as a metal particle. Let us imagine how an electron tunnels from one lead onto the metal particle and then onto the other hand. Suppose the particle is neutral to begin with, to add a charge Q to the particle requires energy $Q^2 / 2C$ where C is the total capacitance between the particle and the rest of the system. When $Q = -Ne$ the peak positions are equally spaced in gate voltage with separation $\Delta V_g = e/C_g$. This results is a direct consequence of the charge quantization. Taking $E (Q) = Q^2 /2C$ means that the energy as a function of Q is a parabola with minimum at Q_o. By varying V_g one can choose any value of Q_o, the charge that would minimize the energy were not the charge quantized. As the real change is quantized only discrete values of energy E are possible. When $Q_o = -Ne$ for which an integer number N of electrons minimizes E the coulomb interaction results in an energy difference $U = e^2 /2C$ for increasing or decreasing N by one. There is thus an energy gap that suppresses charge fluctuations. For all values of Q_o except $Q_o = -(N + 1/2)e$ are degenerate; the charge fluctuations between the two values even at zero temperature. Consequently the energy gap disappears and current can flow. The peaks in conductance are periodic, occurring whenever the average charge on the artificial atom is $Q_o = -(N + 1/2)e$.

The wave functions of electrons are extended to macroscopic distances. Their quantized charge in nanoscale and the degree of localization depends on the transmission of the tunnel barriers. Rc, time constant, for an electron to tunnel off the droplet into the leads be great enough that the energy uncertainty is less than the charging energy. If the tunneling resistance is R, the condition is $Rc > h/U$ or $R > h/e^2$, the fundamental unit of resistance enters in the quantum hall effect. Planck's constant determines whether the charging energy is present or not. The condition is valid at T = 0 independent of C and of the size of the artificial atom.

When electrons are confined to small volumes, the criteria for charge and energy quantization at T = 0 are same. Whereas U is the energy to add an extra electron to the artificial atom there is a typical level spacing ΔE necessary to excite the artificial atom with fixed number of electrons. The levels of the artificial atom are not perfectly sharp but rather have typical width Γ. The level width is caused by lifetime broadening because an electron in a level on the artificial atom can tunnel into the leads. The eigen states of the system are mixtures of localized states on the artificial atom and extended states in the leads.

The energy quantization means that $\Delta E \geq T$. The current through SET for a single quantum level is the charge of the electron divided by the time 't' for an electron in a single quantum state to transverse the artificial atom while in that level. If $(dN/dE)eV_{ds}$ is the number of current carrying channels between the Fermi energy in the source and that in the drain. Thus the current is $I = e/t \ (dN/dE)eV_{ds}$. The width gives the traversal tome, $t = h/t$ and $(dN/dE) = 1/\Delta\varepsilon$ so the condition for energy quantization is $R = Vd_s / I > h / e^2$, the same as for charge quantization.

While the conditions for charge and energy quantization at zero temperature are the same, charge quantization often survives to higher temperatures. The charge quantization can be observed when $kT < U$ but energy quantization requires $kT < \Delta E$. Since $U > \Delta E$ for most SET's made to date, energy quantization is more difficult to observe than charge quantization.

The energy level spectrum is measured directly by observing the tunneling current at fixed V_g as a function of V_{ds}. If $Q_o = -Ne$, V_g increase causes V_{ds} increase. The Fermi level in the sources rises in proportion to V_{ds} relative to the drain so it also rises relative to the energy levels of the artificial atom. Current begins to flow when the E_F of the source is raised just above the first quantized energy level of the atom. As the Fermi energy is raised further higher energy levels in the atom fall below Fermi energy and more current flows because there are additional channels for the electron to use for tunneling onto the artificial atom. One measures the energies by measuring the voltage at which the current increase or the voltage at which there is a peak in the derivative of the current (dI/dV_{ds}).

The diamonds with very low differential conductance are regions where only one charge state is stable. The boundary of the diamonds corresponds to the threshold for changing the charge of the artificial atom. One can overcome the charging energy by changing the source-drain voltage as well as by changing V_g. The diagonal lines outside the diamond correspond to excited energy levels of the artificial atom. In this case, the artificial atom is so small, that $\Delta E \sim U$ so the peak spacing at $V_{ds} = 0$ are far from constant and reflect the shell structure of the artificial atoms. One can ever see the effects of exchange, that is filling follows Hund's rule making certain values of N more stable than others.

There are now three different energy scales U, ΔE and Γ. These are energy to add an electron to the artificial atom, to excite the artificial atom with a fixed number of electrons and the broadening of the artificial atom's energy levels by quantum mechanical tunneling to the leads. Although the coulomb blockade model is often adequate to estimate U, like ΔE and Γ, has quantum mechanical origin. There is one past energy scale that is even smaller than these three. The tunneling results in a kind of chemical bond between the artificial atom and the leads when the artificial atom has a spin. This is the origin of the “kondo effect”. The kondo “bond” is very weak in SET’s. Gold nanoparticles (Au_{55}) with ligand stabilization behave as SET. Tunneling can take place between two of these Au_{55} ligand stabilized clusters when they are in contact with the shell acting as the barrier for the tunneling. SiGe system has a band structure. Admittance spectroscopy and capacitance voltage measurements are used to determine the energy levels of the bound states in QD.

Applications

Infrared transitions involving energy levels of quantum wells have been used for the operation of IR photo detectors. Incoming IR radiation raises electrons to the conduction band and the resulting electric current flow is a measure of incident radiation intensity. A transition from bound state to bound state takes place within the quantum well. A transition to continuum takes place and the continuum begins at the top of the well so the transition is from bound state to quasi bound state. Finally the transition from bound state to mini bond takes place.

Quantum well and quantum wire lasers have been constructed that make use of the laser emission transitions. These devices have conduction electrons for which the confinement and localization in discrete energy levels takes place in one or two dimensions. Hybrid type lasers have been constructed using “dots in a well” such as InAs quantum dots placed in a strained InGaAs quantum well. Another design employed wires or from another point of view they are quantum dots elongated in one direction.

The substrate of a quantum dot laser diode is an n-doped GaAs substrate. The top p-metal layer has a GaAs contact layer immediately below it. Between this contact layer above and the GaAs substrate, there are a pair of 2μm thick $Al_{0.85}$ $Ga_{0.15}$ As cladding or bounding layers that surround a 190nm thick wave guide made of $Al_{0.05}$ $Ga_{0.5}$ As. The wave guide plays the role of conducting the emitted light to the exit ports at the edges of the structure. Centered in the wave guide is a 30nm thick GaAs region and centered in this region are 12 mono layers of $In_{0.5}$ $Ga_{0.5}$ As quantum dots with a density of $1.5\times10^{10}/cm^2$.

Super Conductors and Quantum Members

Super conductivity is present in a material when pairs of electrons condense into a bound state called a “cooper pair” with dimensions of the order of the coherence length ξ. For a semiconductor the penetration depth λ is a measure of the distance that an externally applied magnetic field B_{app} can penetrate into a pure Type I semi conductor. A type II super conductor has both a lower and an upper critical magnetic field $B_{c1} < B_{c2}$. For low applied fields $B_{app} < B_{c1}$, the material acts like a type I super conductor and excludes magnetic flux and at high applied fields $B_{app} > B_{c2}$ the material become normal. In the intermediate range $B_{c1} < B_{app} < B_{c2}$, the magnetic field penetrates into the bulk in the form of tubes of magnetic flux, each of which contain one quantum of flux Φ_0. Each vortex has a core of radius ξ within which the magnetic field is fairly constant, and an outside region of radius, λ where the magnetic field decays with distance r from the core, a decay which has the exponential form exp $(-r/\lambda)$ at large distances away. Vortices may be looked on as the magnetic analog of quantum wires. The transverse dimensions of their core is in the nm range but their length is macroscopic.

1.3 NANOCLUSTERS

Molecules are in general localized entities. In a typical molecular solid, the intermolecular interactions are much weaker than intramolecular. Bonding energies can usually be analyzed as the Sum of individual molecular contributions with small perturbations form the intermolecular forces.

Molecules are of the dimensions of 10nm. One nanometer radius has 25 atoms approximately. Most of the atoms are on the surface of the cluster. However biological molecules like heme molecule has 75 atoms (Fe C_{34} H_{32} O_4N_4).What makes nanomaterials distinct is their size. If the size of the particles is less than the critical characteristic lengths for electrons to conduct, they exhibit different properties.

Atomic clusters are nanosized objects artificially made out of atoms through different techniques. The clusters can have 2 to 1000 of atoms. They have meta-stable structures particular to the energy of their growth and formation. Since clusters are small aggregates of atoms, it is obvious that the number of atoms lying on the surface will be higher than the number of atoms inside i.e. the surface to volume ratio in these aggregates of atoms is quite large compared to the bulk. For a cluster of 1000 atoms, 25% of the atoms lie on the surface. 20% atoms lie for a cluster of 3000 atoms. Even in the very compact structure of 55 atoms has 32 atoms on the surface. Now a new branch of science known as "cluster Science" is being evolved. For small metal clusters the properties are mainly determined by delocalized valence electrons which are considered free electron like. Hence the detailed atomic positions do not play an important role and the ionic cores can be replaced by a uniform positively charged background. This is the Jellium approximation. For finite size clusters, the Jellium model shows that the effects of ionic cores on the valence electrons reduce to an effective potential in which the electrons are confined. This

Fig. 1.7(a) Metal clusters.

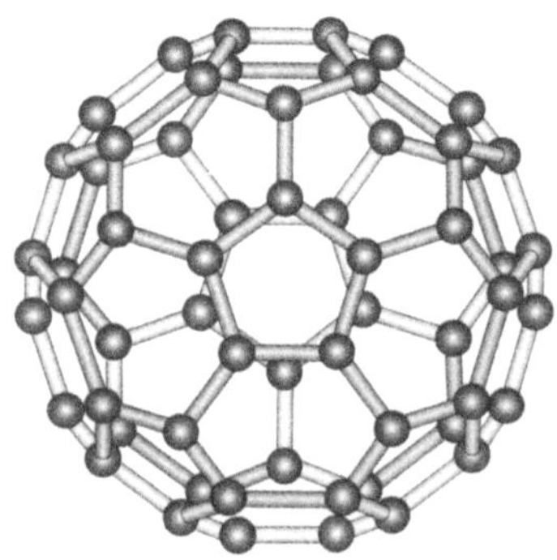

Fig. 1.7(b) Semiconductor clusters.

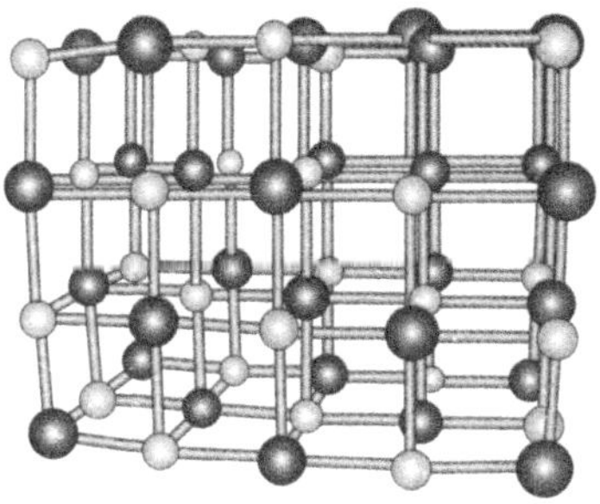

Fig. 1.7(c) Ionic clusters.

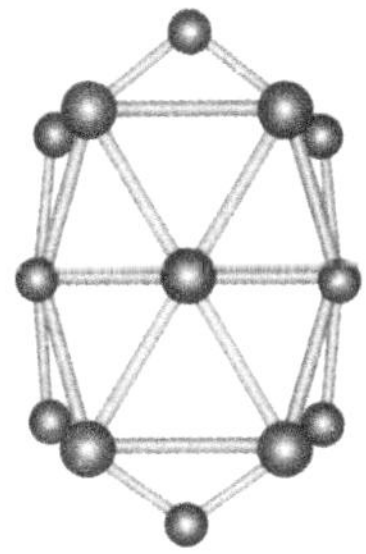

Fig. 1.7(d) Rare gas clusters.

results in the electronic shell model in which the electrons successively fill the different energy levels. **Figure 1.7** presents various nano clusters.

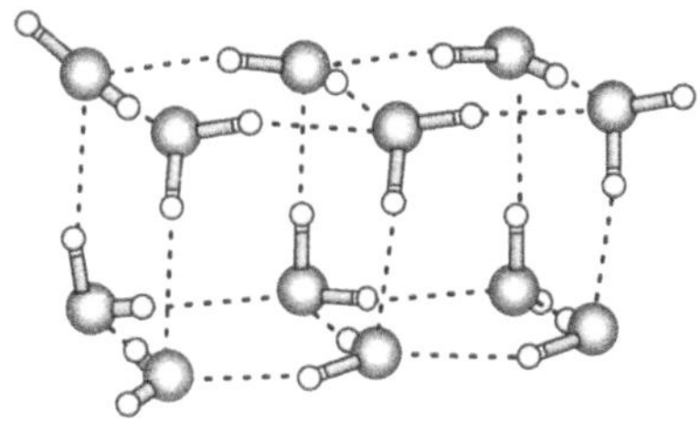

Fig. 1.7(e) Molecular clusters.

Metal nanoclusters can be produced by laser induced evaporation of atoms from the surface of the metal. A high intensity laser beam is incident on a metal rod causing the evaporation. The atoms are then swept away by a burst of Helium and passed through an orifice into a vacuum where the expansion of the gas causes cooling and formation of clusters. These clusters are then ionized by UV radiation and passed into a mass spectrometer that measures their mass: charge ratio. The mass spectra of lead clusters (number of counts vs no. of atoms in cluster) revealed the clusters of 7 and 10 atoms are more likely than others suggested. These clusters are stable.

The ionization potential is the energy necessary to remove the outer electron from the atom. The maximum ionization potentials occur for the rare gas atoms ^{2}He, ^{10}Ne and ^{18}Ar, because their outer most S and P orbitals are filled. More energy is required to remove electrons from filled orbitals than from unfilled orbitals. Ionization potential of sodium clusters as a function of number of atoms in a cluster revealed peaks at 2 and 8 atoms. These numbers are referred to a "Electronic magic numbers". Their existence suggests that clusters of larger clusters can be viewed as superatoms. In the case of larger clusters stability, it is determined by structure and the magic numbers. This is referred to as the "structural magic numbers".

Free Clusters

Cluster of atoms can be viewed as a large atom. The positive nuclear charge of each atom of the cluster is assumed to be uniformly distributed over a sphere the size of the cluster. For small metal clusters the properties are mainly determined by the delocalized valence electrons. Ionic cores can be replaced on a qualitative approximation by a uniform positively charged background. (Jellium Model). For finite size clusters, the ionic cores effect on the valence electrons reduce to an effective potential in which the electrons are confined. Clusters with special 'magic numbers' are predicted to be more stable and to show special features compared to others. **Table 4** compares the energy level scheme for hydrogen atom and the energy level scheme for a spherical positive charge distribution.

TABLE 4. Comparison of the energy levels.

Atoms	Clusters
$1s^2$	$1s^2$
$2s^2$	$1p^6$
$2p^6$	$1d^{10}$
$3s^2$	$2s^2$
	$1f^{14}$
$3p^6$	$2p^6$

Alternative model that has been used to calculate the properties of these clusters is to treat them as molecules and use existing molecular orbitial theories such as density functional theory to calculate their properties. In the quantum theory of the hydrogen atom, the electron circulating about the nucleus is described by a wave. The mathematical function for this wave, called the wave function Ψ. The wave function of the lowest level of the hydrogen atom $\Psi(IS) = A_{exp}\ (-r/p)$ where r = distance of the electron from the nucleus, p = radius of the first Bohr orbit. This comes from solving the Schrodinger equation for an electron having an electron static interaction with a positive nucleus given by e/r.

For H_2^+ ion, molecular orbital theory assumes that the wave function of the electron around the two H nuclei can be described as a linear combination of the wave function of the isolated H atoms. Thus the wavefunctions of the electrtons in the ground state is

$$\Psi = a\ \Psi\ (1)_{1S} + a\ \Psi\ (2)_{1S}$$

The Schrodinger equation is

$$[-(\hbar^2/2m)\nabla^2 - \hbar^2 e^2/r_a - e^2/r_b)\psi] = E\psi$$

The symbol ∇^2 denotes a double differentiation operation. The last two terms in the brackets are the electrostatic attraction of the electron to the positive nuclei which are at distances r_a and r_b from the electron. For the hydrogen molecule, which has two electrons, a term for the electrostatic repulsion of the two electrons would be added. The Schrodinger equation is solved with this layer combination of wave functions.

For simple metal clusters, a transition occurs from the electronic shell structure to a shell structure decided by geometry. Geometry of the nanoparticles are in general the same as the bùlk materials but with some what different lattice parameters. For example, 80nm aluminium particle have shown FCC unit cell. But < 5nm gold particles have icosahedral structure rather than the bulk FCC structure.

Aluminum cluster of 13 atoms have 3 possible arrangements. If one calculates the number of bonds and minimizing the number of atoms on the surface, an aluminium FCC unit cell was found to have different structure. For Bi_N, Pb_N, In_N and Ag_N nanoparticles, deviations from FCC were observed for clusters < 8nm. Indium clusters undergo a change of structure when they are < 5.5nm. Above 6.5nm, a diameter corresponding to about 6000 atoms, the clusters have a FCC centred tetragonal structure with a c/a ratio of 1.075. In a tetragonal unit cell, the edges of the cell are perpendicular, the long axis is denoted by c and the two short axes by a. Below < 6.5nm, the c/a ratio begins to decrease and at 5nm c/a = 1, meaning that the structure is face centered cubic. It is interesting that the structure of isolated nanoparticles may differ from that of ligand stabilized structures. Ligand stabilization refers to associating non metal ion groups with metal atoms/ions. Structural changes cause different properties, (especially electronic structure).

Al_{13} clusters have 2.77 eV as binding energy and Al separation as 2.81 A°. It has an unpaired electron in the outer shell, if one electron is added Al_{13} (–) closes the shell with a significant increase in the binding energy. The molecular orbital approach is able to account for the dependence of the binding energy and ionization energy on the number of atoms in the cluster.

Nanoparticles existing as an isolated entity can be considered. Aluminium nanoparticles are reactive. Air passivated aluminum nanoparticles (80nm) indicated that they have 3.5nm Al_2O_3 layer. If nanoparticles are made from solution, solvent molecules, surfactants, such as oleic acid added to the solution, can adsorb and prevent agglomeration. Metal nanoparticles can be coated with self assembled mono layers. Gold particles can be passivated by using octacylthiol. The long hydrocarbon chain molecule is attached to its end to the gold particle by the thio head group SH. Attractive interactions between the molecules produce a symmetric ordered arrangement of them about the particle. This symmetric arrangement of the molecules around the particle is a key characteristic of the SAM.

When a metal particle having bulk properties is reduced in size to few hundred atoms, the density of states (number of energy levels in a given interval of energy), the top band containing electrons changes dramatically. The continuous density of states in the band is replaced by a set of discrete energy levels which may have energy level spacings larger than the thermal energy, $K_\beta T$ and a gap opens up.

The small cluster is analogous to a molecule having discrete energy levels with bonding and anti bonding or orbitals. Eventually a size is reached where the surfaces of the particles are separated by distances which are in the order of the wavelength of the electrons. This is referred to as quantum size effect. Heisenberg uncertainty principle states that the more an electron is spatially confined, the broader will be not determined so much by the chemical nature of the atoms, but mainly by the dimension of the particle. Quantum size effect can occur for a semiconductor with a wavelength of 1μm where as in a metal it is ~ 0.5nm.

The colour of a material is decided by the wavelength of light that is absorbed by it. Clusters of different sizes will have different electronic structures and different energy level separations. Electronic structure can be studied by UV photo electron spectroscopy. An incident UV photon removes electrons from the outer valence level of the atom. The electrons are counted and their energy are measured.

UV photo electron spectra of the outer levels of copper clusters having 20 and 40 atoms revealed that electronic structure in the valence region varies with the size of the cluster. The energy of the lowest peak is a measure of the electron affinity of the cluster. The electron affinity is defined as the increase in electronic energy of the cluster when an electron is added to it. Reactivity of the nanoclusters depends on size, 3.5nm size gold particles are icosahedral and on Fe_2O_3 substrate they are used as odour eaters for bath rooms.Very small nano particles have all or almost all of their atoms on the surface. Surface atoms have limited mobility and the fluctuations cause different geometries.

In a cluster, the magnetic moment of each atom will interact with the moments of the other atoms and can force all the moments to align in one direction with respect to some symmetry axis of the cluster. The cluster will have a net moment and is said to be magnetised. Stern–Gerlach experinment is used to measure the magnetic moment. The clusters are sent into a region where there is an inhomogenity. Magnetic field which separates the particles according to whether their magnetic moment is up or down. From the beam separation and a knowledge of the strength and gradient of the magnetic field, the magnetic moment can be determined. However for magnetic nanoparticles, the measured magnetic moment is found to be less than the value for a perfect parallel alignment of the moments in the cluster. The atoms of the cluster vibrate and this vibrational energy increases with temperature. These vibrations cause some misalignment of the magnetic moments of the individual atoms of the cluster so that the net magnetic moment of the cluster is less than what it would be if all the atoms had their moments aligned in the same direction. The component of the magnetic moment of all individual clusters will interact with an applied DC magnetic field and is more likely to align parallel than antiparallel to the field. The overall net moment will be lower at higher temperature. It is inversely proportional to the temperature an effect called "super para magnetism". When the interaction energy between the magnetic moment of a cluster and applied magnetic field is greater than the vibrational energy; there is no vibrational averaging. As the clusters rotate, there is some averaging .This is called "Locked moment magnetism". Non magnetic atoms exhibit magnetic moment. For example 80 atoms of Rhenium was non magnetic and below 15 atoms it had magnetic moment rise to 0.8 Bohr magnetons. A transition from bulk to nano size caused changes in melting point. In a cluster < 100 atoms the amount of energy needed to ionize it, that is to remove an electron from the cluster differs from the work function. Physico chemical properties of bulk materials change as the size reduces. From 11 nm as size decreases, the melting point of gold decreases and at a 2 nm size gold has a melting point of $500^{\circ}C$ (**Figure 1.8**). Clusters of gold had been found to have the same melting point only when they contain 1000 atoms. As the gold particles diameter decreased, melting point also decreased to 800 K. A 15nm metal cluster at room temperature follows Ohm's law

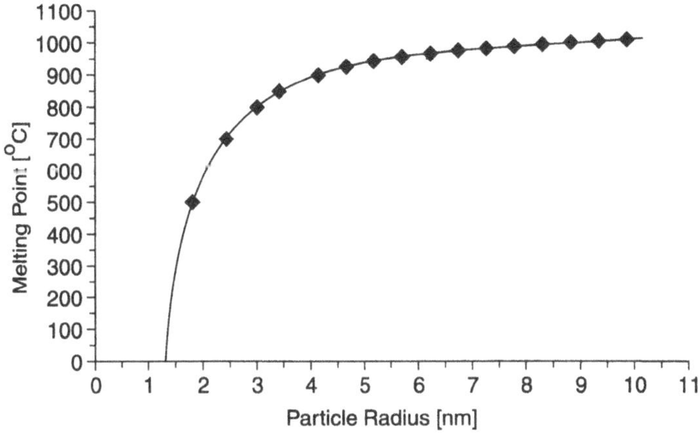

Fig. 1.8 Variation of melting point of gold with size.

whereas a 1.5nm cluster behaves as a Quantum dot. 1.4nm Au_{55} clusters with various ligand shells can be organized in 3 dimensions. Ligand free Au_{55} clusters act as cluster superstructures.

Semiconductor Clusters

In a molecular solid, the intramolecular interactions are much weaker than the intramolecular bonding energies and so the bulk properties of a molecular solid can usually be analyzed as the sum of individual molecular contributions with small perturbations from the inter molecular forces. Such weak intermolecular interactions rarely extend beyond the nearest neighbours and the electronic properties of a molecular crystal are usually independent of the size of the crystal. For a semiconductor crystal, electronic excitation consists of a loose bounded electron hole pair and usually delocalized over a length much larger than the lattice constant. As the diameter of the semiconductor crystallite approaches this exciton Bohr diameter, its electronic properties start to change. This is so called "Quantum size effect."

The most striking property of nanoparticles is the pronounced changes in their optical properties compared to those of bulk material. There is a significant shift in the optical absorption spectra toward the blue (shorter wavelength) as the particle size is reduced. Quantum size effect is a result of size. The smaller the size of a semiconductor, the higher the energy of the excited state. Assuming a "perfect" cluster, the volume-normalized oscillator strength of this excited state should increase with decreasing size. Both effects are manifestations of the quantum size effect. As the cluster properties are intermediate between molecules and bulk semiconductors, the quantum size effect is understood with hybrid molecular and semiconductor principles.

In a bulk semiconductor a bound electron hole pair, called an "exciton" can be produced by a photon having an energy greater than that of the band gap of the material. The photon excites an electron from the filled band to the unfilled band. The result is a hole in the otherwise filled valence bond which corresponds to an electron with an effective positive charge. Because of the coulomb interaction a bound pair called an "exciton" is formed that can move through the lattice. The distance of separation is many lattice parameters. The existence of the exciton has a strong influence on the electronic properties of the semiconductor and its optical absorption.

Cluster size and optical properties are related. Optical bandgap (exciton energy) depends on size. Let us consider polyene, an idealized polyacetylene chain. It has one Pπ orbital. As the infinite chain of carbon atoms separated by distance, and each carrying one Pπ orbital provides a simple one dimensional analogue of the bulk soild. The infinite chain is equivalent to a N-annulene when N is infinite.

In the absence of any interaction, an N-atom polyene has N degenerate orbital Φ_i each with on-site energy (coulomb integral) α, $(\Phi_i / H / \Phi_{j)}$ where H is electronic Hamiltonian. By turning on the interaction between Φ_i and Φ_j represented by the transfer integral, $\beta <\Phi_i / H / \Phi_j>$ the degeneracy is removed. Then the lowest energy orbital is at $\alpha + 2\beta$ and is bonding between all neighbouring atoms. The highest energy orbital is at $\alpha - 2\beta$ and is antibonding between neighbouring atoms. In the middle at energy α, there is a non bonding orbital.

It was shown that as the size of the N-annulene or N-polyene decreases, its energy band becomes discrete and separation between the eigen values and HOMO-LUMO gaps increase. This is in essence the primary part of the quantum size effect, i.e. the increase of the excited state energy with decreasing cluster size.

In a bulk semiconductor, the electron and hole are bound together by a screened coulomb interaction to form a so called Mott-Wannier exciton. The Bohr radius of the exciton is given by

$$a_B = \frac{\hbar^2 \varepsilon}{e^2}[1/m_e + 1/m_h]$$

where ε is the dielectric constant and m_e and m_h are the electron and hole effective mass respectively. This electron-hole interaction is important for quantum size effect. By assuming the energy band to be parabolic near the bandgap (i.e. the effective mass approximation) the size dependent shift (with respect to the bulk bandgap) in the exciton energy of small cluster (cluster radius-exciton radius) can be derived as

$$\Delta E = \frac{\hbar^2 \pi^2}{2R^2}\left[\frac{1}{m_e} + \frac{1}{m_h}\right] - \frac{1.786e^2}{\varepsilon R} - 0.248\, E^*_{RY}$$

where R is the cluster radius and E^*_{RY} is the effective Rydberg energy,

$$E^*_{RY} = \frac{e^4}{2\varepsilon^2 \hbar^2}\left[\frac{1}{m_e} + \frac{1}{m_h}\right]$$

The first term represents the particle in a box quantum localization energy and has a simple $1/R^2$ dependence, the second term represents the Coulomb energy with 1/R dependence and the third term is a result of the spatial correlation effect. This last size independent term is usually small but can become significant for semiconductors with small dielectric constant.

The size of the nano particle and the radius of the orbital of the electron-hole pair affect optical properties. There are two situations called the weak confinement and the strong confinement regimes. In the weak regime, the particle radius is larger than the radius of the electron-hole pair but the range of motion of the exciton is limited which causes a blue shift of the absorption spectrum. When the radius of the particle is smaller, than the orbital radius of the electron-hole pair, the motion of the electron and the hole become independent and the exciton does not exist. The hole and the electron have their own set of energy levels. Here there is also a blue shift and the emergence of a new set of energy levels. In the case of CdSe, the absorption edge, which is due to the band gap, increased with decrease in particle size.

Cluster size and exciton oscillator strength are related. The size dependence of the exciton oscillator strength are being studied. In a bulk semiconductor, the electron and hole are bounded by the screened coulomb interaction with a binding energy of a few to tens of milli eV. This exciton is easily ionized at the thermal energies which accounts for the absence of a strong exciton absorption band in bulk semiconductor at room temperature. By confining the electron and hole in a small cluster, the binding energy and the oscillator strength can increase due to the enhanced spatial overlap between the electron and hole wave functions and the coherent motion of the exciton. This confinement effect is responsible for the "gaint oscillator strength" effect observed for bound exciton in semiconductors, where the electron-hole pair is confined in the potential well established by the shallow defect.

Time resolved laser spectroscopic technique is used to generate surface trapped electron-hole pair on semiconductor clusters. A short laser pulse is used to excite the semiconductor cluster generating initially a bound exciton. This bound exciton is rapidly trapped by the surface to form a trapped electron-hole pair, which subsequently decays with complex kinetics lasting from tens of picoseconds to nanoseconds and beyond. During the lifetime of the bound exciton and the trapped electron-hole pair, their effects on the optical absorption spectrum of the clusters can be probed by a second weak pulse.

In the early time domain (sub picosecond) the presence of the bound exciton can also affect the absorption spectrum. This exciton-exciton interaction in principle can lead to the formation of a biexciton.

The clusters in general are surrounded by dielectrics such as air, polymer, glass or solvent usually with lower refractive indices.When illuminated by light, the field intensity near at and insides the cluster surface can be enhanced considerably compared to the incident intensity because of this boundary established by the different refractive indices. This local enhancement effect can have important consequences on the photo physical and non linear optical properties.

When a laser light passes through silicon or germanium clusters, the clusters fragment. Some of the fissions that have been studied are $Si_{12} + hv > 2\ Si_6$;

$Si_{20} + hv > 2\ Si_{10}$. When the cluster size is greater than 30 atoms, the fragmentation has been observed to occur explosively.

Multiple ionization of clusters causes them to become unstable resulting in very rapid high energy dissociation or explosion. The fragment velocities from this process are very high. This phenomena is called "coulombic explosion". Multiple ionizations of a cluster cause a rapid redistribution of the charges on the atoms of the cluster making each atom more positive. If the strength of the electrostatic repulsion between the atoms is greater than the binding energy between the atoms, the atoms will rapidly fly apart from each other with high velocities. The minimum number of atoms N required for a cluster of charge Q to be stable depends on the kinds of atoms and the nature of the bonding between the atoms of the cluster. **Table 5** gives the smallest size that is stable for doubly charged clusters of different kinds.

Larger crystals are more readily stabilized at higher degrees of ionizations. Clusters of inert gases tend to be larger because their atoms have closed shells that are held together by much weaker forces called Vander Waals forces. The attractive forces between the atoms of the cluster can be overcome by the electrostatic repulsion between the atoms when they become positively charged as a result of photo ionization.

TABLE 5. Smallest obtainable multiply charged clusters.

Atom	**Charge**		
	+2	**+3**	**+4**
Kr	Kr_{73}	—	—
Xe	Xe_{52}	Xe_{14}	Xe_{206}
CO_2	$(CO_2)_{44}$	$(CO_2)_{106}$	$(CO_2)_{216}$
Si	Si_3		
Au	Au_3		
Pb	Pb_7		

Less Common Clusters

Rare gas clusters of krypton and xenon are known. Xenon clusters are formed by adiabatic expansion of a supersonic jet of the gas through a small capillary into a vacuum. The gas is then collected by a mass spectrometer where it is ionized by an electron beam and its mass: charge ratio measured. For the case of Xe, the most stable clusters occur at particles having 13, 19, 25, 55, 71, 87 and 147 atoms. Ar clusters have similar structural magic numbers. The forces that bond inert gas atoms into clusters are weaker than those that bond metals and semiconductor atoms. Two inert gas atoms have an attractive potential known as Vander Waals potential which is effective at relatively large separation of atoms. As the two atoms get much closer together, they repel. The over all interaction potential between two inert gas atoms has the form

$$U\ (R) = B/R^{12} - C/R^{6}$$

where R is the distance of separation. This is known as Lennard-Jones potential and is used to calculate the structure of inert gas clusters. The force between the atoms arising from this potential is a minimum for the equilibrium distance $R_{min} = (2B/C)^{1/6}$ which is attractive for larger separations and repulsive for smaller separations of the atoms. It is weaker than the forces that bind metal and semiconducting atoms into clusters.

Supersonic free jet expansion of helium gas have been studied by mass spectroscopy. The magic numbers are found at clusters size of N = 7,10,14,23,30 for ^{4}He and N = 7,10,14,21,30 for ^{3}He. One of the more unusual properties displayed by clusters is the observation of super fluidity in He clusters having 64 and 128 atoms. Super fluidity is the result of the difference in the behaviour of atomic particles having half integer spin called "Fermions" and particles having integer spin called "Bosons". The difference between them lies in the rules that determine how they occupy the energy levels of a system. Fermions like electrons are only allowed to have 2 particles in each energy level with their spins oppositely aligned. Bosons on the other hand do not have this restriction. This means as the temperature is lowered and more and more of the lower levels become occupied. Bosons can all occupy the lowest level where fermions will be distributed in pairs at the lowest sequence of levels. The case where all the bosons are in the lowest level is known as "Bose-Einstein condensation". When this occurs the wavelength of each Boson is the same as every other and all of the waves are in phase. When Boson condensation occurs in liquid He^4 at the temperature 2.2K, called the lambda point (λ) the liquid He becomes a super fluid and its viscosity drops to zero. In the super fluid state the liquid moves quickly through the tube, and increasing the pressure at one end does not change the velocity. The transition to the super fluid state at 2.2K for liquid He is marked by a discontinuity in the specific heat known as the "Lambda Transition".

Molecular Clusters

Water molecules exist as clusters. The broad Raman spectra of the O–H stretch of the water molecule in the liquid phase at 3200-3600 cm^{-1} has been shown to be due to a number of overlapping peaks arising from both isolated water molecules and water molecules H-bonded unto clusters. The H-atom of one molecule forms a bond with the oxygen atom of another. Increase of temperature split water molecular clusters. The molecular clusters are $(NH_3)_n^+$, $(CO_2)_{44}$ and $(C_4H_8)_{30}$.

Supported Clusters

One of the major challenges in nanotechnology is the long term stabilization of clusters against aggregation and growth by different mechanisms. Nuclei growth can be controlled by precipitation under starving conditions. A large number of nucleation centres are formed by vigorous mixing of the reactant solutions. If concentration is kept small, nuclei growth is stopped due to lack of material. During growth small particles dissolve and are consumed by larger particles. As a result the average nanoparticle size increases with time and the particle concentration decreases. As size increases, solubility decreases. This is Oswald ripening. Stabilizers are added in solution to prevent uncontrollable growth of particles, prevent particle aggregation control growth rate, controls particle size, and allow particle solubility in various solvents. Homogeneous nucleation occurs via a stepwise sequence of bimolecular additions until a nucleus of critical size is obtained. Highly mono disperse nanoparticles are formed if the process of nucleation and growth can be successfully separated. Nucleus radius is calculated as

$$\Delta G = 4\pi\sigma\, [n^2\, (2r^3/3r^*)]$$

where r = nucleus radius; = critical nucleus radius; σ = surface tension.

$$\Delta G_{(nucleus)} = n(\Delta G_{\text{formation, bulk}} - \Delta G_{\text{formation, free atom}}) + \sigma A$$

where A = particle surface area.

A commonly employed route for stabilizing nanoparticles is the use of a protective layer of organic molecules which can efficiently adsorb on the surface of the clusters. Self assembled monolayer formation is common. In this process since long chain organic molecules such as thiols, amines and carboxylic acids are arranged on the cluster surface in a periodic manner using effective inter chain Vander Waals interaction. The clusters are prevented from agglomeration and degradation. However these monolayer protected nanoparticles have limited stability in air possibly due to the diffusion of oxygen through pinholes and other defects present in the monolayer causing degradation upon storage. For example, dodecane thiol capped gold clusters have been found to degrade after two weeks upon storage in air while preparation of copper clusters always gives contamination due to oxides. Encapsulation of the nanoclusters in a passive medium can solve most of the above problems of degradation/stability. And hence a variety of inorganic materials including zeolites, glasses gels and polymers are used to stabilise.

Nanoclusters on Polymers

Polymers are particularly popular as they possess several unique features compared to inorganic silicates and other materials for encapsulation. They either as passive or active matrix can provide and stable environment for regular dispersion of shape controlled particles where the flexibility and ready processing can be used for device fabrication. Polymers provide a transparent and non conducting support and non lineat optical properties of clusters are studied. Suitable control of hydrophobic and hydrophilic groups on polymer surface modulation of exact location of nanoparticles on the surface is achieved.

Polymer and nanoclusters interactions can be tailored. For example alkyl chain surrounding gold nanoparticle was made to interact with alkyl chains of polymer. The amide or carboxyl group containing polymers was made to interact with a gold cluster containing amino thiophenol. To organize clusters hydrophobic, hydrogen bonding and π cation interactions were used.

Nanocluster Organization

Electrostatic interaction plays an important role. The charge distribution on polymer surface is by attaching specific acidic or basic functional groups such as $-COOH$, SO_3H, $-NH_2$ etc. can electrostatically interact with the terminal part of the ligands. 2-5nm size nanoparticles of Au, Pt and Ag with surface carboxyl group were organized on poly (amidoamine) dendrimers using in situ synthesis.

Covalent interaction arises mainly when a covalent bond is formed between the surface functionality present on the clusters and suitable groups on the polymer surface. In this mode of organization the directional nature of the bonding can be used to fabricate cluster arrays. The polymer surface can be modified using different functional groups such as –SH, –CN, –S–S etc., which are known to form covalent bonds with both metallic as well as semiconducting clusters. Covalently bound thiol capped polystyrene macro molecules were used to organize functionalized gold nanoparticles. Hydrophobic interactions are used to organize a monolayer protected gold clusters (MPC) on poly ethytane or poly styrene surface.

Selection of Clusters

Selection of clusters especially their size, shape and distribution plays a crucial role in determining the effectiveness of organization on the polymer surface. Clusters can be mainly divided into two categories viz., metallic and semiconducting clusters.

Metallic nanoclusters like Au, Ag, Cu, Pt etc., in combination with diverse polymeric systems have been used. If the size of the these metallic particles is lesser than the electron mean free path, it gives surface plasmon absorption and this gives a characteristic colour to metallic nanoparticles in colloidal solution. Magnetic nanoclusters such as Fe, Co, Ni are also of immense technological importance due to their preferred orientation.

The organization of desired semiconducting clusters (Quantum dots) on a special polymeric materials can be used for a broad range of applications. As the diameter of the semiconducting clusters approaches Bohr diameter, their electronic efficiency of optical excitation to create electron/hole pair as well as consequent radiative decay can be indirectly controlled by Quantum-dot polymer interactions. When semiconducting clusters are encapsulated with polymers having lower refractive index the local field intensity of the clusters gets enhanced to the incident intensity after illumination and such dielectric confinement helps to study the optical properties of these systems.

Limitations

Though the problems of uncontrolled aggregation of nanoparticles has been overcome through spontaneous adsorption of capping of agents on the surface of the clusters, these methods still require the large concentrations of capping agents which can interfere with the subsequent functionalization upon organization of nanoclusters on polymer surface by preferred path. The enhanced mobility may exist on the polymer surface. Another limitation involves the variation of the polymer molecular weight which affects reproducibility. Metal nanoparticles organized on polymer support exhibit interesting optical, electrical and catalytic properties. Pt, Pd supported functional polymers can have superior catalytic property of semiconductor nanorod – polymer hybrid solar cells where high electron affinity semiconductors like

CdSe coupled with poly (3–hexyl thiophene) are used. Gold colloids covalently derivatised with hydrophilic polymer bearing terminal disulphide groups are compatible with biological systems.

Glass Supported Metal Clusters

Dielectric matrices embedded with nanoscale metal clusters are the potential materials for use as optical wave length devices. Embedded metal clusters in silicate glasses exhibit a large third order non-linear susceptibility near the surface plasmon resonance frequency with a picosecond response time. Heterogeneous materials consisting of metal nanoclusters in dielectric matrices are synthesized by various techniques like low energy ion beam mixing, sol-gel, direct metal implantation, light ion irradiation of ion exchanged glasses, thermal annealing of ion exchanged glasses and co-deposition.

Metal nanoclusters are formed in ceramics and semiconductors by thermal treatment and ion irradiation. Silver has been deposited on zirconia powder pellet by vacuum evaporation and annealed at 800°C for 8 to 10 hours to enable silver atoms to diffuse into the pores of the pellet. Usually silver cluster size ranges from 2 to 200nm. Vacuum deposited CdS thin films on glass has been irradiated by ions like N^+ and Ar^+ and Cd metal nanoclusters of ~2nm size embedded in CdS matrix. Silver is usually deposited on silica surface. In the ion beam mixing of Ag atoms with in silica to a depth of 400nm by Ar^+ irradiation, excess Ag was dissolved out of the surface and was thermally annealed at various temperatures to form Ag nanoclusters. Silver can also be incorporated into soda lime glass by ion exchange process to a depth of ~5000nm and thermally annealed to form silver nanoclusters.

1.4 CARBON NANOSTRUCTURES

Carbon contains six electrons which are distributed over the lowest energy levels of the carbon atom. The lowest energy level 1s with the quantum number N = 1 contains electrons with oppositely paired electron spins. The electron charge distribution in an s state is spherically symmetric about the nucleus. The next 4 electrons are in N = 2 energy state, one is a spherically symmetric s orbital and three in p_x p_y and p_z orbitals, which have the very directed charge distributions oriented perpendicular to each other. This outer s orbital together with the three p orbitals form the chemical bonds of carbon with other atoms. The charge distribution associated with these orbitals mixes (or overlaps) with the charge distribution of each other atom being bonded to carbon.

In the carbon atom, the energy separation of between the 2s level and 2p levels is very small and this allows and admixture of the 2s wavefunction with one or more of the $2p_i$ wave functions. The unnormalized wave function Ψ in a valence state can be designated as $\Psi = s + \lambda p$. Where p indicates an admixture of p_i orbitals. With this hybridization the directions of the p lobes and the angles between them changes. The angles will depend on the relative admixture λ of the P states with the S state. Diagonal sp hybridization has $\lambda = 1$ with a bond angle 180°. Trigonal sp^2 hybridization has $\lambda = \sqrt{2}$ with a bond angle 120°; Tetrahedral sp^3 has $\lambda = \sqrt{3}$ with a bond angle 109° 28′.

Diamond consists of carbon atoms that are tetrahedrally bonded to each other through sp^3 hybrid bonds that form a three dimensional network. Graphite has a layered structure with each layer, called a graphite sheet formed from hexagons of carbon atom bound together by sp^2 hybrid bonds that make 120° angles with each other. Each carbon atom has 3 nearest neighbour carbons in the planar layer. The hexagonal sheets are held together by weaker Vander Waals forces (**Figures 1.9 and 1.10**).

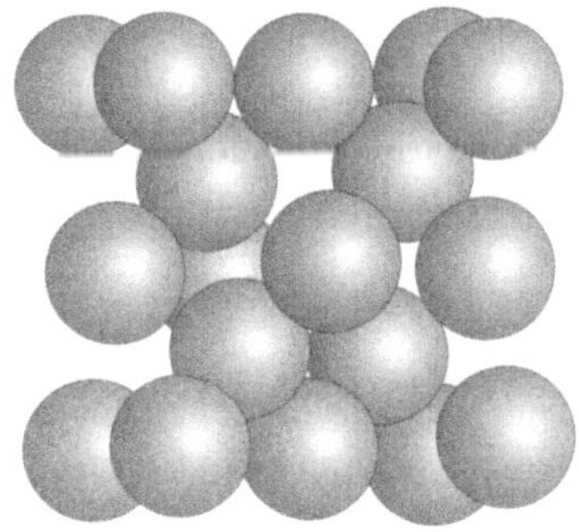

Fig. 1.9 Diamond unit cell.

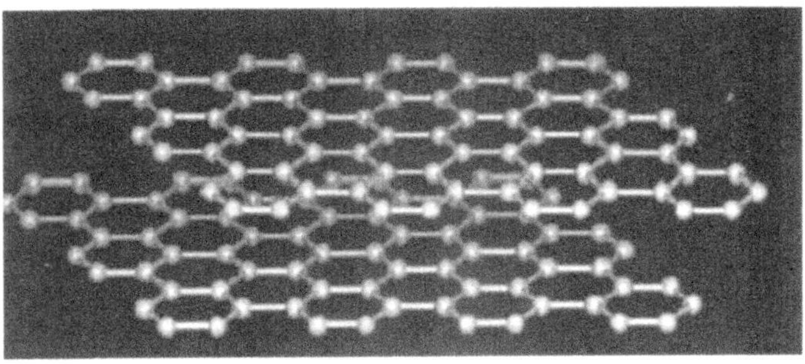

Fig. 1.10 Graphic structure.

In 1964 Phil Eaton of the University of Chicago synthesized a square carbon molecule, C_8H_8, called Cubane. In 1983, Paquette of Ohio state University synthesized a $C_{20}H_{20}$ molecule having a dodecahedron shape formed by joining carbon pentagons and having C – C bond angles ranging from 108° to 110°. The idea of making carbon bond with various bond angles inspired the research on carbon clusters.

Carbon Clusters

A Nobel prize work on the laser evaporation of a carbon substrate in a pulse of He gas can be used to make carbon structures. The neutral cluster beam is photo ionised by a UV laser and analysed by a mass spectrometer. Mass spectra revealed for the number of atoms $N < 30$, there are clusters of every N, although some are more predominant than others. Calculations of the structure of small clusters by molecular orbital theory show that the clusters have linear or closed non polar mono cyclic geometries. The linear structures that have SP hybridization occur when N is odd and the closed structures form when N is even. The open structures with 3, 11, 15, 19 & 23 carbons have the usual bond angles and more prominent and more stable. The closed structures have angles between the carbon bonds that differ from those predicted by conventional hybridization concept. A very large mass peak at 60 was noticed.

Although a Soccer ball like molecule consisting of 60 carbon atoms with the chemical formula C_{60} had been envisioned by theory, in 1990, an article in "Nature" came out with the existence of 60 carbon atoms. Later an experiment was carried out at Rice university by Richard Smalley and Harlod Kroto. A graphite disk was heated by a high intensity laser beam that produces a hot vapour of carbon. A burst of He gas then swept the vapour of carbon out through an opening where the beam expands. The expansion cooled the atoms and they condense into clusters. This cooled clusters beam was then narrowed by a skimmer and fed into a mass spectrometer which is a device designed to measure the mass of molecules in the clusters. A mass number of 720 that would consist of 60 carbon atoms, each of mass 12 was seen. This was the evidence for a C_{60} molecule. The C_{60} molecule was named after the architect R.Buckminister Fuller. The name *Buckminister fullerene* was shortened to fullerene. It has 12 pentagonal (5 sided) and 20 hexagonal (6 sided) faces symmetrically arrayed to form a molecular ball (shape similar to Soccer ball).

These ball – like molecule bind each other in the solid state to form a crystal lattice having a FCC structure. Each C_{60} molecule separated from its nearest neighbour by 1nm and are held together by weak forces called Vander Waals forces. C_{60} is soluble in benzene and single crystals can be grown by evaporation of benzene. 26% of the volume of the unit cell is empty so alkali atoms can easily fit into empty spaces. In the tetrahedral site the alkali atom has 4 surrounding C_{60} molecules. When the radius of the dopant alkali atom increases, the cubic C_{60} lattice expands and the super conducting transition temperature increased (**Figure 1.11**).

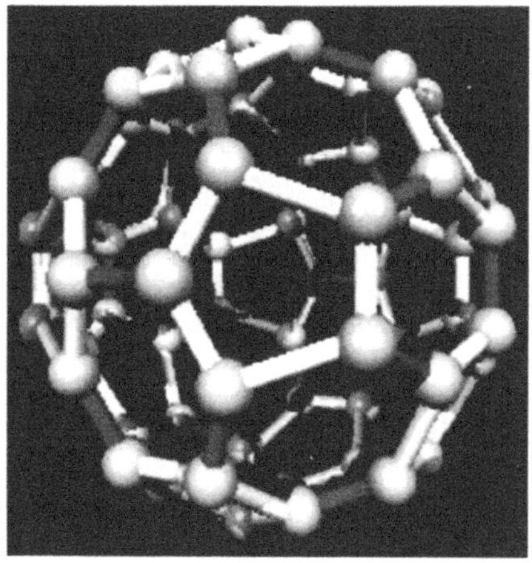

Fig. 1.11 Fullerenes (C-60).

Larger fullerenes such as C_{70}, C_{76}, C_{80} and C_{84} are also known. By gas phase dissociation of $C_{20}HBr_{13}$, a C_{20} dodecahedral carbon molecule was synthesized $C_{36}H_4$ was made by pulsed laser ablation of graphite. A solid phase of C_{22} has been identified in which the lattice consists of C_{20} molecules bonded together by an intermediate carbon atom. When appropriately doped these smaller fullerenes conduct as super conductors.

Carbon Nanotubes

It is a member of fullerene structural family. Nanotubes are cylindrical with at least one end typically capped with a hemisphere of the bulky ball structure. There are two main types of nanotubes: Single walled nanotubes and multiwalled nanotubes. Most SWNT's have a diameter of close to 1nm with a tube length that can be many thousands of times larger. SWNT's with length upto orders of centimeters are produced (Figure 1.12) The structure of a SWNT can be conceptualized by wrapping a one-atom thick layer of graphite called grapheme into a seamless cylinder. The way the

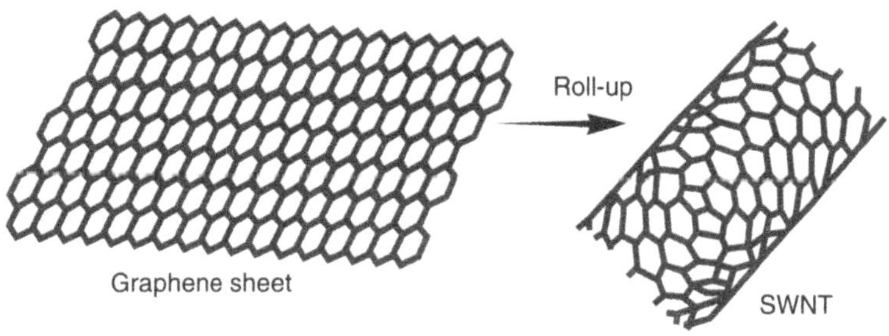

Fig. 1.12 Single all carbon nanotube.

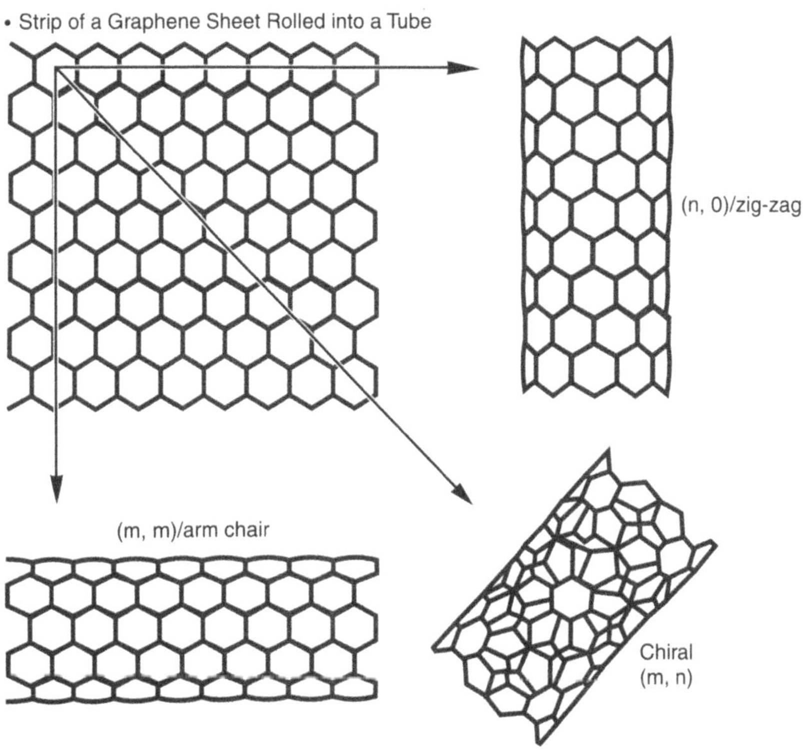

Fig. 1.13 Various manifestations of carbon nanotubes.

graphene sheet is wrapped is represented by a pair of indices (n, m) called the chiral vector. The integers n and m denote the number of unit vectors along two directions in the honeycomb crystal lattice of graphene. If $m=0$, the nanotubes are called "zig-zag". If $n=m$, the nanotubes are called "armchair". Otherwise, they are called "chiral". Multiwalled (MWNT) consist of multiple layers of graphite rolled in on themselves to form a tube shape. Although carbon nanotubes are not actually made by rolling graphite sheets it is possible to explain the different structures by considering of the way graphite sheets may be rolled into tubes (**Figure 1.12 and Figure 1.13**).

There are two models which can be used to describe the structures of multi walled nanotubes. In the Russian doll model, sheets of graphite are arranged in concentric cylinders. In the parchment model, a single sheet of graphite is rolled in around itself resembling a scroll of parchment or a rolled up newspaper. Fullerite is a highly incompressible nanotube form. Polymerised single walled nanotubes are a class of fullerites and comparable to diamond in terms of hardness. A nanotorus is a carbon nanotube bent into a torus (donut shape). It has 1000 times magnetic moment larger than previously expected for certain specified radii. **Table 6** presents the chronological developments in nanotubes.

TABLE 6. Chronological developments in Nanotubes.

Year	Development
1985	Discovery of fullerene
1991	Discovery of nanotubes in Japan
1993	Single wall carbon nanotubes-developed by IBM and NEC, Japan
1995	Nanotransistor by Delft and IBM
2001	Semiconductor surfaces from nanotubes by IBM
2002	Multiwalled nanotubes
2003	High purity nanotubes with metallic property
2004	Replacement of tungsten filament in a light
	Bulb with a carbon nanotube; 4cm long SWNT
2005	T shaped nanotubes by university of California carbon; nanodiode by GEC, USA.
	Nanotube sheet 5 × 100 cm.
	Multiwalled nanotube of 1mm.
2006	IBM built an electronic circuit around a CNT

Properties : Structure of carbon nanotubes falls into SWNT and MWNT. Multiwall (nested cylinder) and single wall types are known. Depending on the roll over these can be classified as 3 different distinct varieties viz. arm chair, zig-zag and chiral types. A nanotube can be formed when a graphite sheet is rolled up about the axis T. The C_h vector is called the circumferential vector and it is at right angles to T. Three examples of nanotube structures constructed by rolling the graphite sheet about the T vector having different orientations in the graphite. When T is parallel to the C–C bonds of the carbon hexagons, the structure is obtained is "armchair". "zig-zag" and "chiral" structures are formed by rolling about a T vector having different orientations in the graphite plane but not paralled to C–C-bonds. Looking down the tube of the chiral structures, one would see a spiraling row of carbon atoms. Generally nanotubes are closed at both ends which involves the introduction of a pentagonal topological arrangement on each end of the cylinder. The tubes are essentially

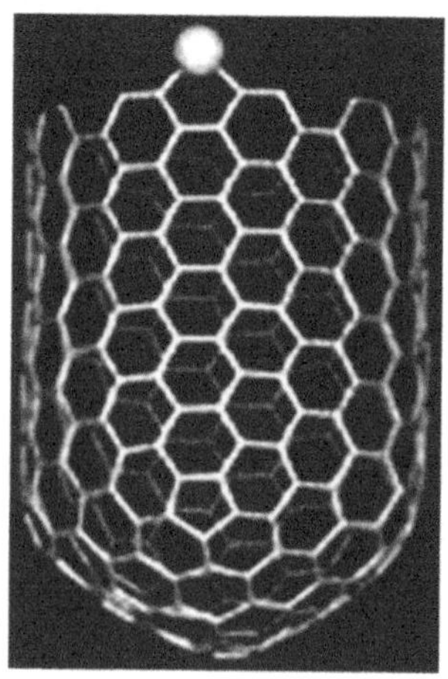

Fig. 1.14 (10, 10) carbon nanotube.

cylinders with each end attached to half of a large fullerene like structure. **Figure 1.14** presents the (10,10) carbon nanotubes.

Carbon nanotubes have diameter as small as 0.1nm. Their electronic property have been predicted to vary sensitively with the diameter and helicity of the tube lattice. Slight changes in these parameters causes a shift from metallic to semimetallic state. i.e., the same carbon atoms in the same planar graphite lattice can be have so differently properties of single wall tubes depend on the diameter and helicity of the tubes. Armchair tubes are found to be metallic where as the zig-zag and chiral tubes can be both metallic as well as semiconducting. As per the chiral reactor notation (n,n) defines the arm chair, both (n,o) and (o,m) defines the zig-zag (n,m). Theory predicts that whenever (n-m) is divisible by 3 the tube is metallic, other wise semiconducting.

Scanning tunneling microscopy has been used to investigate the electronic structure of carbon nanotubes. In this measurement the position of the STM tip is fixed above the nanotube and the voltage V between the tip and the sample is swept while the tunneling current I is monitored. The measure of the local electronic density of states. The density of states is a measure of how close together the energy levels are to each other. STM data plotted as the differential conductance (dI/dV)/ (I/V).vs the applied voltage is used. The energy gap in materials at voltages where very little current is observed can be seen. For semiconductor it is 0.7eV. At higher energies sharp peaks are observed in the density of states (Van Hove singularities) and are characteristic of low dimensional conductors. The peaks would appear at the bottom and top of a number of sub bands. (**Figure 1.15**).

If the electron wavelength is not a multiple of the circumference of the tube, it will destructively interfere with itself and therefore only electron wavelengths that

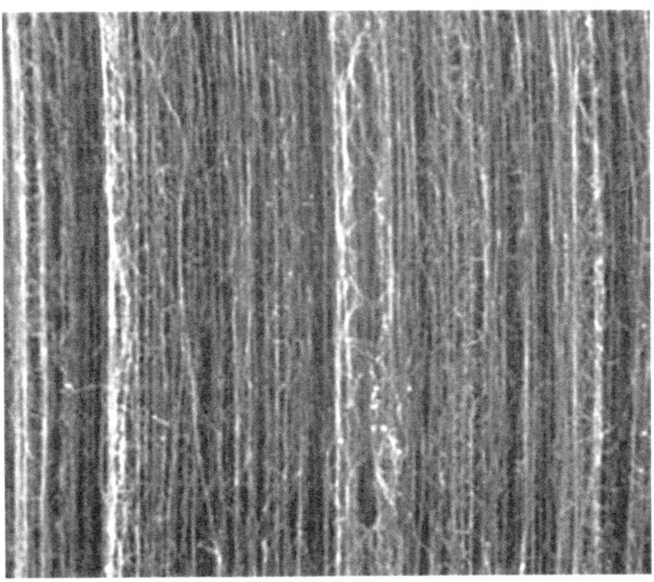

Fig. 1.15 Multiwall carbon nanotubes.

are integer multiples of the circumference of the tube are allowed. This severely limits the number of energy states available for conduction around the cylinder. The dominant remaining conduction path is along the axis of the tubes, making carbon nanotubes function as one dimensional quantum wires. The electronic states of the tubes do not form a single wide electronic energy band, but instead split into one dimensional sub bands. These states can be modeled by a potential well having a depth equal to the length of the nanotubes.

Electron transport has been measured on individual single walled carbon nanotubes, at $T = 10^{-3}K$ on a single metallic nanotube lying across two metal measurements. The steps occur at voltages which depend on the voltage applied to a third electrode that is electrostatically coupled to the nanotubes. This resembles a field effect transistor made from a carbon nanotubes. The step like features in the I –V curve are due to single electron tunneling and resonant tunneling through single molecular orbital. Single electron tunneling occurs when the capacitance of the nanotube is so small that adding a single electron requires an electrostatic charging energy greater than the thermal energy. Electron transport is blocked at low voltages, which is called "Coulomb blockade". By increasing the gate voltage, electrons are added to the tube one by one. Electron transport in the tube occurs by means of electron tunneling through discrete electron states. The current at each step is caused by one additional molecular orbital. This means that the electrons in the nanotube are not strongly localized but rather are spatially extended over a large distance along the tube. Generally in one dimensional systems the presence of a defect will cause a localization of the electrons. However a defect in a nanotubes will not cause localization because the effect will be averaged over the entire circumference because of the doughnut shape of the electron wave function. In the metallic state, the conductivity of the nanotubes is high due to very few defects to scatter electrons. High currents do not heat the tubes in the same way that they heat copper wires. Nanotubes have a very high thermal conductivity almost a factor of 2 greater than that of diamond. Carbon multi wall tubes could pass a very high current density up to $10^{10}A/cm^2$. CNT does not influence the electrical transport property CNT – FET shows a p–type semiconductor. To perform the logic functions, both p–type and n–type CNT – FET's are required. The electrical properties can be modified by chemical doping with potassium and annealing in O_2. Hydrogen functionalization of CNT leads to the transformation of metallic (narrow gap semiconducting) to semi-conducting. This is due to the C–H bond inducing sp^3 hybridization and thus removing the π and π^* bands near the Fermi level opening the energy gap.

Carbon nanotubes have normal modes of vibration. One mode, m labelled A— involves an 'in and out' oscillation of the diameter of the tube. Another mode, the E_{2g} mode, involves a squashing of the tube where it squeezes down in one direction and expands in the perpendicular direction essentially oscillating between a sphere and an ellipse. The frequencies of these two modes are Raman active and depend on the radius of the tube. The dependence of this frequency on the radius is now routinely used to measure the radius of nanotubes.

Carbon nanotubes are strong. They have Young's Moduli ranging form 1.28 to 1.8 Tpa. One tera pascal is a pressure very close to 10^7 times atmospheric pressure. (For steel the Young's moduli is 0.21 Tpa). They are very stiff and hard to bend. If bent they are very resilient. They buckle like straws but do not break and can straightened back without any damage. They have so few defects in the structure of the walls if bent the almost hexagonal carbon rings in the walls change in structure but do not break. This is a unique result of the fact that the C–C bonds are sp^2 hybrids and these sp^2 bonds can re hybridize as they are bent. The degree of change and the amount of s–p and mixture both depend on the degree of bending of the bonds. The tensile strength of CNT is 45 billion pascals (high strength steel 2 billion pascals). Nested nanotubes also have improved mechanical properties but they are not as good as their SWNT. MWNT of 200nm diameter have a tensile strength of 0.007 Tpa and a modulus of 0.6 TPa.

Applications

Carbon nanotubes are known to be the best available field emitters. Their high aspect ratio, high chemical stability, high thermal conductivity and high mechanical strength are advantageous. They could be used not only as a switching device and interconnect wires but also as a memory device. This is done by fabricating a non-voltatile memory based kin CNT–FET's and oxide-nitride-oxide (ONO) storage nodes. CNT based memory devices are expected to have high performance.

A plastic composite of CNT could provide light weight shielding material for electromagnetic radiation. For computer inter connects carbon nanotubes of 2nm diameter are used. They serve as heat sinks. For switching device, CNT are used. 10^{12} switches could fit on a $1cm^2$ chip. Pentium chips have 10^8 switches. The switching rates are 100 times faster. CNTs are used in Fuel cells. Single walled CNT is the form paper are negative electrode in a KOH solution. A counter electrode is $Ni(OH)_2$. Chiral semiconducting CNT are used as NO_2 sensors. Nested CNT with Ru metal bonded to the outside has a catalytic effect for the hydrogenation of cinnamaldehyde. CNT are used for improving mechanical properties. Adding to polypropylene, 11.5% by weight of nested CNT having an 0.2μm diameter approximately doubled the tensile strength of the polypropylene. 5% by volume of nanotubes in aluminium increased the tensile strength by a factor of 2. The CNT to form strong bonds with iron suggesting the possibility of adding CNT to steel to improve strength. Kelly-Tyson equation suggests the tensile strength of steel could be increased by 7 times if the steel had 30% by weight of oriented carbon nanotubes in it. Carbon nanotubes additionally is used to produce nanowires of gold or ZnO. These nanowires in turn are used to cast nanotubes of other chemicals such as GaN. Gallium nitride nanotubes are hydrophilic unlike CNT. Carbon nanotube fibers and films are used. Spinning fibers od SWNT with polyvinyl alcohol were made. Transparent carbon nanosheets $1/1000^{th}$ of thickness of the human hair were developed. Carbon tube nanofibers are electrically conductive and strong.

1.5 INORGANIC NANOTUBES AND NANOWIRES

Several layered inorganic compounds possess structures comparable to the structure of graphite, the metal dichaclogenides (sulphides, selenides and tellurides) halides (Cl^-, Br^-, I^-), oxides numerous ternary (quaternary) compounds. Atoms are also arranged so that they may look like tubes wound over nanotubes (**Figure 1.16**).

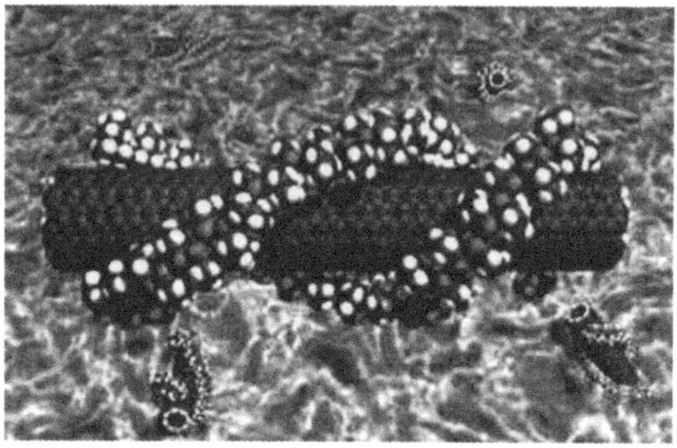

Fig. 1.16 Nanotubes are builts by arranging several atoms in a closed pattern to form an unbelievably small tube.

The metal dichalcogenides, MX_2 (M = Mo, W, Nb, Hf, X = S, Se) contain a metal layer sandwiched between two chaliogen layers with the metal in a trigonal pyramid or octahedral coordination mode. The MX_2 layers are stacked along the c-direction in ABAB fashion. The MX_2 layers are analogous to the single graphene sheets in the graphite structure. However in contrast to graphite, each molecular sheet consists of layers of different atoms chemically bonded together. When viewed parallel to the—C-axis, the layers show the presence of dangling bonds due to the absence of an X or M at the edges. Such unsaturated bonds at the edges of layers also occur in graphite. The dichalcogenide layers are unstable towards bending and have a high propensity to roll into curved structures. If the dimensions of the dichalcogenide layers are small, then they form hollow, closed clusters designed as Inorganic fullerene like (IF) structures.

Before the discovery of the carbon tubes, as early as 1979 "Layered transition metal chalcogenides" (LTMC) was recognized. Mo and W dichalcogenides are capable of forming nanotubes. Closed fullerene type structures are formed from this. The dichalcogenide structures contain concentrically nested fullerene cylinders with a less regular structures than in the carbon nanotubes. Accordingly MX_2 nanotubes have varying wall thickness and contain some amorphous material on the exterior of the tubes. Nearly defect free MX_2 nanotubes are rigid as a consequence of their structure and do not permit plastic deformation.

Folding may arise due :

1. Stoichiometric LTMC chains and layers such as those of TiS_2 possess have an inherent ability to bend and fold.

2. The existence of alternative coordination and therefore of stoichiometry in the LTMC.

3. A change in the stoichiometry within the material would give rise to closed rings.

Dichalcogenides of many of the group 4 or 5 metals have layered structures suitable for forming nanotubes.

Asbestos mineral (chrysotil) also have fibrous nature which is determined by the tubular structure of the fused tetrahedral and octahedral layers. Mesoporous silica of 2-20nm is also known. Silica nanotubes were first produced by as a spin off product during the synthesis of spherical silica particles by the hydrolysis of tetraethyl orthosilicate (TEOS) in a mixture of water, NH_3, ethanol and D,L-tartatic acid. Boron nitride (BN) crystallizes in a graphite like structure and can be simply viewed as replacing a C–C pair in the graphene sheet with the iso electronic B–N pair. It is an ideal precursor for the formation of BN nanotubes. Replacement of the C–C pairs in the hexagonal network of graphite leads to the formation of a wide array of two dimensional phases that can form hollow cage structures and nanotubes. The possibility of replacing C–C pairs by B–N pairs in the hollow of C_{60} is also tried.

Properties and Applications

$NbSe_2$ nanotubes are found to be metallic at ordinary temperatures. BN tubes are insulators with a wide band gap of 5.5eV. The chalcogenide nanotubes with an average ~6A° Vander Waals gap between the layers are used for H_2 storage. Multiwalled BN nanotubes have a capacity to uptake 1-8 to 2.6 wt % of the under ~10Mpa at 30°C. Though smaller than the reported for CNT's it is useful. MoS_2 nanotubes could be electrochemically charged and discharged with a capacity of 260m Ahg^{-1} at 20°C corresponding to a formula of $H_{1.24}MoS_2$. The high storage capacity is believed to be due to the enhanced electrochemical catalytic activity of the highly nanoporous structure. This may find wide applications in high energy batteries.

BN nanotubes are insulators and used for encapsulating conducting materials like metallic wires. Filled BN nanotubes are expected to be useful in nanoscale electronic devices and for the preparation of nanostructured ceramics.

The most likely applications of the Mo and W chalcogenide nanotubes is as solid lubricants. The hollow nanoparticles of WS_2 exhibit tribological properties. Tribological properties of 2H – MoS_2 and WS_2 powder can be attributed to the weak Vander Waals forces between the layers which allow easy shear of the films w.r.t. each other. The mechanism in the WS_2 nanostructures is somewhat different and the better tribological properties may arise from the rolling friction allowed by the round shape of the nanostructures. Open tipped MoS_2 nanotubes are used as catalysts for CO to CH_4.

Nanowires

One dimensional (1D) nanostructures such as nanowires, nanorods and nanobelts provide good models to investigate the dependence of electronic transport, optical, mechanical and other properties on size confinement and dimensionality. Nanowires are likely to play a crucial role as interconnects and active components in nanoscale devices. The synthesis of nanowires with controlled composition, size, purity and crystallinity requires a proper understanding of the nucleation and growth processes at the nm regime.

While comparing with bulk materials, low dimensional nanoscale materials with large surface area and possible quantum confinement effect exhibit distinct electric, optical chemical edge of the Si nanowires was strongly blue shifted from the bulk indirect band gap of 1.1eV. They observed sharp discrete absorbance features and relatively strong "band edge' photoluminescence. They attribute this to a quantum confinement effect as well as to lattice orientation of the nanowires. Indium phosphide nanowires exhibited striking anisotropy in the PL intensity recorded parallel and perpendicular to the long axis of the nanowire. This anisotropy is used to create polarization—sensitive nanoscale photo electors which may be useful in optical switches, high resolution detectors and integrated photonic devices. ZnO nanowires exhibit room temperature ultraviolet lasing. They are used as natural resonance cavities.

Semiconducting nanowires are used as building blocks for assembling a range of nanodevices including FET's, p-n diodies, bipolar junction transistors and complementary inverters. The nanotubes and nanowires with sharp tips are promising materials for applications as cold cathode field emission devices. Field emission characteristics of both β-SiC and Si nanowires exhibit robost field emission with turn-on fields of 15 & 20° V/μm respectively, with a c.d. of 0.01ma/cm^2 comparable with diamond and CNT.

Metallic Nanorods

Fundamentally, the mean free path of an electron in a metal at room temperature is 10–100nm, and as the metallic particle shrinks to this dimension, unusual effects might be expected.

The aspect ratio of a solid is defined as its length divided by its width, therefore spheres have an aspect ratio of 1. A nanorod may be defined as object with an aspect ratio between 1–20 with the short dimension on the 10–100nm scale. A nanowire by this definition has an aspect ratio > 20.

Depending on the aspect ratio, the extinction spectra (the combination visible absorption and scattering) of Ag and Au nanoparticles are tunable through out the visible depending on the aspect ratio. For nanorods and nanowires, the plasmon band of the metal is split in two : the longitudinal plasmon bond, corresponding to light absorption and scattering along the long axis of the particle and the transverse plasmon band corresponding to light absorption and scattering along the short axis of the particle. Metallic nanorods have tunable optical properties through out the visible and into the IR portions of the electromagnetic spectrum and they are predicted to have enhanced SERS activity compared to spheres. As the size of integrated circuit elements shrinks to the ~ 100nm scale and below, metallic and semiconducting nanowires of controllable size and ability to be positioned need to be developed. The magnetic domain size in magnetic nanoparticles is ~10nm and control of crystal structure and size of the relevant metals and alloys (e.g. Fe, Co) and this scale are key to advances in magnetic data storage. Hybrid data storage schemes that rely on melting nanorods to nanospheres in polymer matrices with subsequent detection of the altered optical properties for "reading" and "writing" are being developed.

All of the promise of the future technology based on nm scale inorganic solids relies on the production of nanoparticles of controlled size, shape and crystal structure. This ultimately requires that these nanoparticles have to be rationally linked to make a working device.

Nanostructured Ordered/Disordered Materials

The solid nanoparticles can be disordered with respect to each other where their symmetry axes are randomly oriented and their spatial positions display no symmetry. The particles can be ordered in lattice arrays displaying symmetry.

Disordered Structures

Fracture occurs due to the existence of cracks in the material. A 'crack' is essentially a region of a material where there is no bonding between adjacent atoms of the lattice. If such a material is subjected to tension, the crack interrupts the flow of stress. The stress accumulates at the bond at the end of the crack making the stress at that bond very high, perhaps exceeding the bond strength. This results in a breaking of the bond at the end of the crack and lengthening the crack. A crack provides a mechanism where by a weak external force can break stronger bonds one by one. Another kind of mechanical failure is the brittle-to-ductile transition where the stress-strain curve deviates from linearity. Young modulus is the factor relating stress and strain. It is the slope of the stress-strain curve in the linear region. The larger the value of Young's modulus, the less elastic the material. For nano grain Fe, below >> 20nm Young's modulus begins to decrease from its value in conventional grain sized materials. The yield strength σ_y of a conventional grain sized material is related to the grain size by the Hall-Petch equation,

$$\sigma_y = \sigma_o + K/\sqrt{d}$$

where σ_o is the frictional stress opposing dislocation movement, K is a constant and d is the grain size in micro meters. Assuming this equation is valid for nano sized grains, a bulk materials having a 50nm grain size would have a yield strength of 4.14GPa. The reason for this increase in yield strength with smaller grain size is that materials having smaller grains have more grain boundaries, blocking dislocation movement. Deviations from Hall-Petch equation is observed for materials < 20nm. The deviations involve no dependence on particle size (zero slope) to decrease in yield strength with particle size (negative slope). It is believed that conventional dislocation–based deformation is not possible in bulk nano structural materials with sizes < 30nm because mobile dislocations are unlikely to occur. Examination of small grained bulk nano materials by TEM during deformation does not show any evidence for mobile dislocations.

Most bulk nanostructured materials are quite brittle and display reduced ductility under tension, typically having elongations of a few % for grain sizes <30nm. For example conventional coarse grained annealed poly crystalline copper is very ductile having elongations of upto 60%. Measurements in samples with grains sizes <30nm yield elongations no more than 5%. Most of the measurements have been performed on consolidated particular samples which have large residues stress and flaws due to imperfect particle bonding which restricts dislocations movement. Nano structured copper prepared by electro deposition displays almost no residual stress and has elongations upto 30%.

Nanostructured multi layers are made by sputter deposition, chemical vapour deposition and electrochemical deposition. They have very large interface area densities. This means that the density of atoms on the planar boundary between two layers is very high. For example a cm^2 of a 1μm thick multi layer film having layers of 2nm thickness has an interface area of 1000 cm^2. These layered materials have very hardness which depends on the thickness of the layer and good wear resistance. Hardness is measured using an indentation lead depth sensing apparatus which is known as "Nanoindentor". For TiN/NbN nano multilayered structures, as the layers get thinner in the nm range there is a significant enhancement of the hardness until 30nm where it appears to level off and become constant. Multi layers in which the alternating layers have different crystal structures are found to be even harder. In this case dislocations moved less easily between the layers and essentially became confined in the layers resulting in an increased hardness.

Consider a bulk nanostructured gold nanoparticles. They are connected by long molecules. Their network is made by taking the gold particles in the form of an aerosol spray and subjecting them to a fine mist of a thiol such as dodecanethiol RSH, where R is $C_{12}H_{25}$. These alkyl thiols have an end group — SH that can attach to a CH_3 and a methylene chain 8-12 units long that provides steric repulsion between the chains. The chain like molecules radiate out form the particles. The encapsulated gold particles are stable in aliphatic solvents such as hexane. However the addition of a small amount of dithiol to the solution causes the formation of a 3 dimensional cluster network that precipitates out of the solution. Clusters of particles can also be deposited on flat surfaces once the colloidal solution of encapsulated nanoparticles has been formed. Nanoparticles are prevented from aggregation by a process known as stabilization. **Figure 1.17** shows a nanoparticle stabilized by thiols.

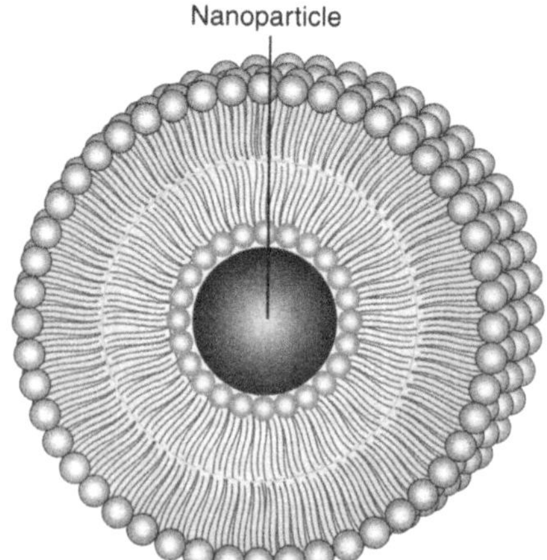

Fig. 1.17 Stabilized nanoparticle.

The nanostructured alloy $Fe_{73}B_{13}Si_9$ has been found to have enhanced resistance to oxidation at temperatures between 200 to 400°C. The enhanced resistance is attributed to the larger number of interface boundaries, and due to the fact that atom diffusion occurs faster in nanostructured materials at high temperatures. In this material the Si atoms in the FeSi phase segregate to interface boundaries where they can then diffuse to the surface of the sample. At the surface the Si interacts with the O_2 in the air to form a protective layer of SiO_2 which hinders further oxidation. The melting point of nanostructured materials is also affected by grain size 4nm Indium particles has its m.pt lowered by 110K. Glasses containing a low concentration of dispersed nanoclusters display a variety of unusual optical properties. The effect of the size of gold nanoparticles on the optical absorption properties of an SiO_2 glass in the visible region is well known. The peak of the optical absorption shifts to shorter wavelengths when the nanoparticles size decreases from 80 to 20nm. The spectrum is due to plasma absorption in the metal nanoparticles. At very high frequencies the conduction electrons in a metal behave like a plasma, that is, likely an electrically neutral ionized gas, in which the negative charges are the mobile electron and the positive charges reside on the stationary background atoms. Provided the clusters are smaller than the wavelength of the incident visible light and are well dispersed so that they can be considered non interacting, the electromagnetic wave of the light beam causes an oscillation of the electron plasma that results in absorption of the light.

Ordered Structures

12 atom boron cluster is an example for "Natural nanocrystals" i.e., one with 20 faces. There are a number of crystalline phases of solid boron containing the B_{12} cluster as a subunit. One such phase with tetrahedral symmetry has 50 boron atoms in the unit cell comprising $4B_{12}$ icosahedra bonded to each other by an intermediary boron atom that links the clusters. Another phase consists of B_{12} icosahedral clusters arranged in a hexagonal array.

There is a possibility of designing a new class of solid materials whose constituent units are not atoms or ions, but rather clusters of atoms. To the cluster $Al_{12}C$ has 40 electrons when a carbon is added. It is stable. The FCC $Al_{12}C$ is a semiconductor and has a band gap of 0.05eV. If KAl_{13} clusters are considered, the electron affinity of Al_{13} is close to that of Cl, it may be possible for this cluster to form a structure similar to KCl.

Zeolites such as the cubic mineral, faujasite, (Na_2Ca) $(Al_2SiO_4)O_{12}.8H_2O$ are porous materials in which the pores have a regular arrangement in space. The pores are large enough to accommodate small clusters. The clusters are stabilized in the pores by weak Vander Waals interactions between the cluster and the zeolite. The pores are filled by injection of the guest material in the molten state. It is possible to make lower dimensional nanostructured solids by this approach using a zeolite material such as mordenite. The mordenite has long parallel channels running through it with a diameter of 0.6nm. Selenium can be incorporated into these channels forming chains of single atoms. A trigonal crystal of Se also has parallel chains but the chains are sufficients close together so that there is an interaction between them. In the mordenite this interaction is reduced significantly and the electronic structure is different from that of a selenium crystal. This causes the optical absorption spectra of the Se crystal and the Se in mordenite to differ.

A two – phase water toluene reduction of $AuCl_4^-$ by sodium borohydride in the presence of an alkane thiol solution produces gold nanoparticles. When the material is examined by X-ray diffraction, it showed in addition to the diffuse peaks from the planes of gold atoms in the nanoparticles Au_m, a sequence of sharp peaks at low scattering angles indicating that the Au nanoparticles had formed a giant three dimensional lattice in the SR matrix.

Super lattices of Ag nanoparticles are electrically neutral and ordered arrangement. When the size < 3nm, the order of the de Broglie wavelength of the conduction electrons, the metal clusters may exhibit novel electronic properties. They display very large optical polarizabilities and non-linear electrical conductance having small thermal activation energies. Coulomb blockade and coulomb stair case current-voltage curves are obtained.

Colloids are small spherical particles of 10–100nm size suspended in a liquid. To prevent agglomeration electrostatic charges are introduced. Another method is to attach soluble polymer chains to the particles, in effect producing a dense brush with flexible bristles around the particles occupy over 50% of the volume of the material the particles begin to order into lattices. The structure of the lattices is generally hexagonal close packed FCC or BCC. By adjusting the fraction between the FCC and BCC can be induced.

A photonic crystal consists of a lattice of dielectric particles with separations on the order of the wavelength of visible light. Such crystals have interesting optical properties. In 1987 Yablonovitch and John proposed the idea of building a lattice with separations such that light could undergo Bragg reflections in the lattice. For Visible light this requires a lattice dimension of 0.5μm or 500nm. This is 1000 times layer than the spacing in atomic crystals but still 100 times smaller than the thickness of a human hair. Such crystals have to be artificially fabricated by methods like electron beam lithography or X-ray a periodic array of dielectric particles having separations of the order of 500nm. The fabrication consisted of covering a block of a dielectric material with a mask consisting of an ordered array of holes and drilling through these holes in the block on 3 perpendicular facets. A technique of stacking micro machined wafers of silicon at consisphotonic structures. Another approach is to build the lattice out of isolated dielectric materials that are not in contact.

1.6 ORGANIC NANOCRYSTALS

π conjugated organic compounds are prepared by a reprecipitation method that involves pouring a rich solution into a poor solvent (usually water) with vigorous agitation such as sonication. After the pouring there are initially widely scattered droplets that gather into dispersed clusters that undergo a process of nucleation and growth until they finally produce the nanocrystals. As these crystallites form, they scatter light and the intensity of the scattered light. I_S (t) relative to the incident light intensity I_o after a time, t, has elapsed can be used to monitor the rate at which the growth takes place. The time dependence of $(I_S(t)/I_o)$ follows the expression $[1\text{-exp}\,(\alpha_{app}\, t)^2]$ where the growth rate constant α_{app} depends on the temperature. The linearity of the plot which is called an Arrehenius plot and provides the activation energy for the crystal growth process, and for perylene nanocrystals, it is 68KJ/mol. The activation energy is the minimum amount of energy that must be supplied by for the nanocrystals, to form. The size of the crystallite can be regulated by varying concentration, temperature and mixing procedure and also by the use of surfactants that modify the surface of the particles or reduce the surface tension of the solution.

Many π conjugated organics in single crystal and thin film configurations exhibit 3rd order non-linear optical properties that makes them useful for converting visible light frequencies to higher frequencies in the UV region of the spectrum. They also have an ultra fast response time so they can be used for switching light beams in which a strong applied electric field activates their biregringence (double refraction). Single crystals containing individual conjugated polymer chains that stretch across the entire length of the crystalline should constitute favourable material for molecular device design.

Polydiacetylene types of compounds are also of sizes in nm. Diacetylene polymers have the capability of forming perfect crystals in the solid state. These marterials have been used in non-linear optics. Small nanocrystals of DCHD exhibit quantum size effects. **Table 7** summarises organic nanocrystals and their size.

TABLE 7. Organic nanocrystals.

Type	Material	Crystal size
	Anthracene	150 nm – 1 mm
	PIC	200 – 300 nm
π conjugated	Perylene	50 – 200 nm
	Fullerene	200 nm
	4 BCMU	200 nm – 1 μm
	DCHD	15 – 1000 nm
Poly diacetylene	DCHD.fiber	1000 nm – 60 nm dia
	14–8.ADA	15 – 200 nm

Supra Molecular Structures

Supra molecular structures are large molecules formed by grouping or bonding together several small molecules. Usually the process of self assembly is used to fabricate these molecules. A square supra molecular structure is made by starting with an angular subunit and combining it with either a linear subunit or another angular subunit. We are familiar with the construction of a tree how one trunk forms several large branches and so on. The roots of the tree exhibit the same branched mode of growth. This type of architecture is a fractal one, associated with a space whose dimensions are not an integer such as 2 or 3 but rather the dimensionality is a fraction. There are molecules dendrimers or cascade molecules. Individual dendritic compounds can be linked together to form larger structures called "supra molecular dendrimers" (**Figure 1.18**) The supra molecular structures-dendrimers, which can be characterized by open or globular structures with central core or hyper branched structures without central core. This new type of highly branched macromolecular structures is suitable for the construction of the biosensor (**Figure 1.19**). Dendritic molecules has an architecture which is a fractal one, associated with a space whose dimensions are not an integer such as 2 or 3 but rather the dimensionality is a fraction. There are molecules called "Dendrimers or cascade molecules".

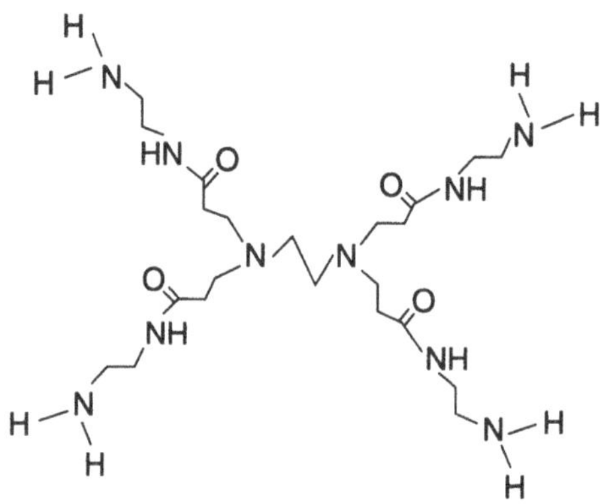

Fig. 1.18 Supra molecular structure—single unit.

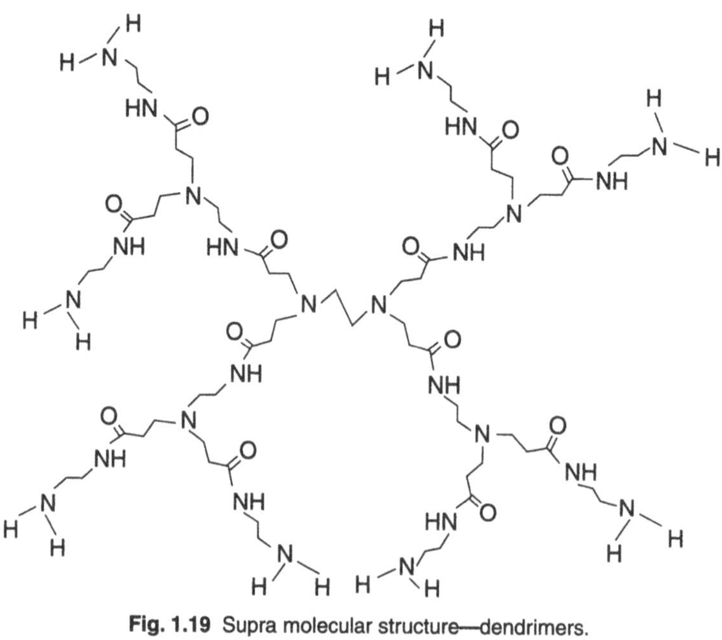

Fig. 1.19 Supra molecular structure—dendrimers.

Diamine ($NH_2 - NH_2$), m-phenylene diamine or 2,6-diaminopyridine may be used as a starting material. The starting compound ($NH_2 - NH_2$) was reacted with vinyl cyanide to replace the hydrogens of the amino groups with cyano groups and form the double cyanide derivative $(CN)_2 N - X - N (CN)_2$. This derivative compound of the dendrimer iteration sequence.

One can use crystallographic data for polyamino amine dendrimer (PAMAM) to estimate size. The average transverse unit cell dimensions are $a_o = 0.415nm$ and $b_o = 4.96nm$ corresponding to the cross sectional area $a_o b_o = 0.206nm^2$. The length of the unit cell C_o is proportional to the number of carbon atoms n and has the average value 0.137nm/carbon. Each PAMAM monomer has 5 carbon atoms and 2 nitrogens corresponding to n = 7 which gives a length of 0.96nm. Taking into account bending at the splitting points or bifurcations, the radius increases by perhaps 1-3nm per generation which gives a total of 13nm for 10 generations. Thus dendrimers of this type have sizes of nanoparticles.

The dendrimers which have the number of terminal groups doubles at each branching point is referred to as "divergent growth", and the process is termed divergent synthesis. Each main branching complex emanating from the cone is called a wedge, which means that the polyamine dendrimer has 2 wedges and the poly amino amine has 3 wedges. Thus a typical dendrimer consists of a central core plus 2,3 or more wedges, each of which ends with an outer region or periphery consisting of terminal groups.

Dendrimers consisting of terminal group containing catalytic sites are sometimes referred to as dendralysts. This dendrimer has arylnickel complexes bound at the periphery of the carbosilane skeleton. It is used to catalyse the particular chemical reaction called the Kharash addition of CCl_4 to the polymer precursor material methyl methacrylates.

Individual dendritic compounds can be linked together to form larger structures called "supramolecular dendrimers". For example a dendrimer in which the core consists of a central benzene ring associated with a complicated aromatic, polycyclic complex compounds attached above it plus 2 branched wedges. Many dendrimers tenaciously hold back some solvent and some of them can trap molecules such as radicals, charged moieties and dyes. When molecules of different sizes are trapped, they can be selectively released by gradual hydrolysis of the dendrimer outer and middle layers. Dendrimers of this type can be useful for prolonging the lifetime of unstable chemical molecules. The torroidal shaped β–cyclodextrin molecule has a hydrophobic central cavity that has a range of inner radii from 0.5 to 0.8nm depending on the number of D–Glycosyl units in the cyclic polymer chain. The number can vary from 6 to 8 or more. The molecule is able to trap energetic molecules such as trinitroazedine and remove them from water effluents so that they will contaminate the environment.

1.7 NANOFIBERS

Botanists identify this term, "Nanofiber" with elongated thick walled cells that give strength and support to plant tissue. The word "fiber" came from Latin word "fibra". A nanofiber is a nanomaterial in view of its diameter and is considered a nanostructure if filled with nanoparticles to form composite nanofibers. The fibers have a high degree of orientation along the fiber axis. The fibers have increased surface area to volume, decrease in pore size and molecular orientation.

The molecular structure of a nanofiber is characterized by FTIR and NMR. If two materials are blended together for nanofiber fabrication, not only the structure of the two can be detected but also intermolecular reaction can be tested. When collagen and PEO blend are used for electrospinning nanofibers, the NMR spectrum showed a new phase structure which is caused by the hydrogen bond formation between the either oxygen of PEO and the protons of the amino and hydroxyl groups in collagen.

During spinning, the polymer molecule has no time to crystallize and can only have an amorphous super molecular structure. The transition points of the polymers also change. Nanofibers processed through self assembly from a polymer solution is directly observed under TEM. Pore mesh structure in a nanofibrous mesh is observed by SEM, TEM and AFM. The porosity of electrospun nanofibrous structure is 91.63% indicating that it is a highly porous one. The total pore volume is 9.69 ml/g, the total pore area is 23.54m^2/gm and the pore diameters ranged from 2 to 456 μm are observed.

The surface properties of nanofibers include surface chemical properties, surface compact morphology, topography and surface roughness. Surface chemical properties can be determined by XPS, water contact angle measurement, and FTIR-ATR analysis. AFM is used to measure the roughness of fibers. The roughness value is the arithmetic average of the deviations of height from the central horizontal plane given in terms of mV of the measurement current. Wet spun PU fiber (febrile) with an average diameter of 100nm has a roughness of 26 while disordered has 35.6. Electro spun disordered PU of 1nm average diameter has 74.0. Dynamic moisture vapour permeation (DMPC) cell is used to measure both the moisture vapour transport and the air permeability of continuous films, fabrics, coated textiles. Tensile properties of a nanofibrous mesh can be measured by conventional mechanical testing methods. PLLA nanofibrous mesh has a stress-strain curve similar of that of skin. Cantilever technique is also used to measure the tenacity of a single polymer ultra fine fiber. A cantilever consisting of a 30μm–glass fiber is glued at one end onto a microscope slide and a 15–nylon fiber is attached to at the free end of the glass fiber. The electrospun test fiber is glued with epoxy resin to the free end of the nylon fiber. A part of the same fiber is cut and deposited on a SEM specimen for diameter measurement using SEM. As the sample fiber is stretched with a computer controlled Instron Model 5569, the deflection of the cantilever was measured under light microscopy using a calibrated eyepiece. A chart was used to convert deflection into actual values of fiber tenacity. The elongation to break of the electrospan poly acrylonitrile (PAN) fibers was estimated. The electrospan PAN fibers with a diameter of 1.25μm and length of 10mm exhibited failure at 0.4mm deflection at 41f mg of force and the resulting tenacity was 2.9 gld. The mean elongation at break of thc same fibre was 190% with a standard deviation of 16% for nanotubes with parallel structure. A tensile measurement is being made by pulling long (~2mm) ropes containing tens thousands of the aligned nanotubes with a specially designed stress-strain puller.

The biodegradability of nanofibrous mesh is expected to be different from the biodegradability of the other structural scaffolds. The nanofibrous scaffold can be placed in PBS at 37°C in an incubator or implanted into animals. The biodegradability can be measured from the mass loss and morphology change over a period of time. Cell proferation on nanofibrous scaffolds is being tested by seeding cells on the nanofiber membrane.

Applications

Biomedical applications : Polymeric nanofibers are well in vascular prostheses and breast prostheses. Protein nanofibers are deposited as a thin porous film onto a prosthetic device for implantation on the central nervous system. The film with a gradient fibrous structure is expected to efficiently reduce the stiffness mismatch. The stress concentration at the tissue / device interface, occurs and prevent the fracture or fatigue failure of the device after implantation. Silk, kertain, collagen Viral spike protein tubulin and actin (proteins) cellulose, chitin and mucin are characterized by well organized fibrous structures ranging from nm to mm scale. To make them bio-compatible, they are designed three dimensional scaffolds of matrices that provide temporary templates. Biodegradable scaffolds from nanoscale polymeric fibers may hold the key in adjusting the degradation rate of a specified biomaterial in vivo. There is a method of fabricating biocompatible and bioabsorbable foam which has a gradient in composition and microstructure. A fibrous layer attached on the foam surface for better control surface properties of the structure is known. Nanofibrous structured scaffolds for engineering cartilage tissue is based on the theory that cells attach and organize well around fibers with diameters smaller than the diameter of the cells. Synthetic biodegradable polymers such as (lactic acid), poly (glycolic acid) are made into small fibers.

Drug delivery with polymer nanofibers is based on the principles that dissolution rate of a particle drug increases with surface area of both the drug and the corresponding carrier if needed. Bioabsorbable nanofiber membranes are used for the prevention of surgery induced adhesions.

Polymer nanofibers can also be used for treatment of wound or burn in the human skin and for haemostatic devices. Superfine fibers of biodegradable polymers can be directly sprayed / spin into the injured location of skin with the aid of an electrical field to form a fibrous mat dressing. The fibrous mat dressing can help the formation of normal skin growth resulting from usual treatments. Non-woven nanofibrous membranes for wound dressing can be usually made with pore size ranging from 500nm to 1μm to protect the wound from bacterial penetration via aerosol particle capture mechanisms. The high surface area of nanofibrous membranes ranging from 5 to 100 m^2/g is extremely efficient for fluid absorption and dermal delivery.

Polymer nanofibers are being tried as cosmetic skin care masks for the treatment of skin healing, skin cleansing and other therapeutic or medical properties with/without various additives. These nanofibrous mask with very small interstices and high surface area will facilitate for greater utilization and speed the rat of transfer of additives to the skin for the fullest use of potential of the additive. Such a cosmetic skin mask of electrospun nanofibers is applied directly to the skin for healing.

Filtration applications : Nanofibers for application in pulse clean catridges for dust collection and in cabin air filtration of mining vehicles are known. Polymer nanofibers can be electrostatically charged to modify the ability of electrostatic attraction of particles without increase in pressure drop to further improve filtration efficiency. Electrospinning process thus integrate the spinning and charging of polymers into nanofibers in a single step. Insoluble linear poly (ethylenimine) nanofibers have a greater surface area capable of neutralization of chemical agents without impedance of air and water vapour permeability to the clothing. Electrospun nanofibers offer minimum impedance to moisture vapour diffusion and maximum efficiency in trapping aerosol particles.

Engineering applications : Polybenzimidazole (PBI) nanofibrous non woven mats with average fiber diameter of 300nm can reinforce epoxy resin and improve its Young's modus, E, fracture toughness K_{1c} and fracture energy, G_{1c}. PB1 also provides reinforcement for styrene butadiene rubber. The higher surface to volume ratio may improve the inter laminar toughness of pratical composites. Electrospun nylon 4,6 nanofibers of 30–20nm diameter are used as reinforcement in epoxy. The resulting composite is transparent due to the fiber being smaller than the wavelength of visible light. Polyimide – single wall carbon nanotubes film is made by the foam of nanofibrous film is made by electrospinning to explore a potential application for spacecraft. Carbon nanofibers for composite applications can also be manufactured from their precursors of polymer nanofibers. Compared to carbon nanotubes produced by vapour growth technique, converting polymer nanofibers such as polyacrylonitrile (PAN) and mesophase pitch to carbon nanofibers by stabilization and carbonization can produce continuous, uniform, solid carbon nanofibers.

Electrical and optical applications : Conducting polymers such as polyaniline and its blends can be spinned to produce polymer nanofibers. Conducting nanofibers may be used in fabricating small electronic devices such as Schottky junctions, sensors and actuators. Conducting nanofibers are used in batteries, electrostatic dissipation, corrosion protection electromagnetic interference shielding and photovoltaic devices.

Functional applications : Nanofibers produced from polymers with piezoelectric properties produce piezoelectric nanofibrous devices. Poly (lactic-co-glycolic) acid (PLGA) nanofiber films are employed as a sensor. Fluorescent polymeric nanofibers are used. Thin film sensors to detect ferric and mercuric ions are known. A singe nanofiber coated with two metals at different segments (thermocouple) will create a junction to detect inflammation of coronary arteries with extremely fast response time. Nanothermo couples can be inserted into a cell to monitor the metabolic activities at various locations with in the cell. Multiple nanothermo couples can be circum ferentially mounted on a catheter balloon to allow mapping of the arterial wall temperature.

CHAPTER 2

Synthesis of Nanomaterials

2.1 SYNTHESIS OF METAL COLLOIDS

Nanoparticulate metal colloids are generally defined as isolable particles between 1 and 50nm in size that are prevented from agglomerating by protecting shells. Depending on the protection shell used they can be dispersed in water or organic solvents. Some of the major methods of synthesis are presented.

Chemical Methods

Nanostructured metal colloids are obtained by both the so called "top down" and "bottom up" methods. A typical 'top down' method for example involves the mechanical grinding of bulk metals and subsequent stabilization of the resulting nanosized metal particles by the addition of colloidal protective agents. The 'bottom up' methods of wet chemical reduction of metal salts, electrochemical pathways or the controlled decomposition of metastable organo metallic compounds. A large variety of stabilizers e.g., donor ligands, polymers and surfactants are used to control the growth of the primarily formed nanoclusters and to prevent them from agglomerating. The chemical reduction of transition metal salts in the presence of stabilizing agents to generate zero valent metal colloids in aqueous or organic was known since 1857. The first reproducible standard recipes for the preparation of metal colloids (e.g. for 20nm gold by reduction of [$AuCl_4^-$] with sodium citrate) were known based on nucleation, growth and agglomeration of nanoclusters is known.

The metal salt is reduced to give zero valent metal atoms in the embryonic stage of nucleation. They can colloide in solution with further metal ions, metal atoms or clusters to form an irreversible 'seed' of stable metal nuclei. Depending on the difference of the redox potentials between the metal salt and the reducing agent applied and the strength of the metal-metal bonds, the diameter of the "seed" nuclei can be well below 1nm.

Nanostructured colloidal metals require protective agents for stabilization and to prevent agglomeration. The two basic modes of stabilization are electrostatic and steric. Electrostatic stabilization involves the coulombic repulsion between the particles caused by the electrical double layer formed by ions adsorbed at the particle surface (e.g. sodium citrate) and the corresponding counterions. For example, gold sols are prepared by the reduction of [Au Cl_4^-] with sodium citrate by coordinating sterically demanding organic molecules that act as protective shields on the metallic surface steric stabilization. In this way nanometallic cores are separated from each other and agglomeration is prevented. In general lipophilic protective agents media while hydrophilic agents yield water soluble colloids. For example in Pd, organosols stabilized by tetra alkyl ammonium halides, the metal core is protected by a monolayer of the surfactant coat.

Metal hydrosols are stabilized by Zwitter ionic surfactants which are able to self aggregate and are enclosed in organic double layers. TEM graphs reveal the colloidal Pt particles (~2.8nm) surrounded by a double layer zone of the Zwitter ionic carboxy betaine (3–5nm). The hydrophilic head group of the betaine interacts with the charged metal surface and the lipophilic tail is associated with the tail of a second surfactant molecule resulting in the formation of a hydrophilic outer sphere. Pt or Pt / Au particles can be hosted in the hydrophobic holes of nonionic surfactants e.g. Polyethylene monolaurate.

Reducing Agents

To make silver nanoparticles, stronger reducing agents produce smaller nuclei in the 'seed'. During the so called 'ripening' process these nuclei grow to yield colloidal metal articles in the size range of 1–50nm which have a narrow size distribution. It was assumed that the mechanism for the particle formation is an agglomeration of zero valent nuclei in the 'seed' or alternatively collisions of already formed nuclei with reduced metal atoms. The stepwise reductive formation of Ag_3^+ and Ag_4^+ clusters by spectroscopic methods is known. An autocatalytic pathway is involved in which metal ions are adsorbed and successively reduced at the zero copper protected by cationic surfactants (NR_4^+) is studied by X-ray absorption spectra which indicated the formation of Cu^+ prior to the nucleating of the particles. The size of the resulting metal colloid is determined by the relative rates of nucleation and particle growth although the process taking place during nucleation and particle growth cannot be analyzed separately.

The salt reduction method has the advantage that in the liquid phase it is reproducible and it allows colloidal nanoparticles with a narrow size distribution to be prepared on the multi gram scale. The classical Faraday route via the reduction of [$AuCl^{4-}$] with sodium citrate, for example, is still used to prepare standard 20nm gold sols for histological staining applications. Wet chemical reduction procedures are used for transition metals with different types of stabilizers. $Au_{55}(pph_3)_{12}Cl_6$ (1.4nm), a full shell (magic number) nanocluster stabilized by phosphine ligands is prepared. Clusters of Au_{55} are uniformly formed when a stream of B_2H_6 is carefully introduced into a Au^{III} ion solution. The diborate 'route' for $M_{55}L_{12}Cl_n$ nanoclusters is also known. The phosphane ligands may be exchanged in the Au_{55} nanoclusters quantitatively using silsequioxanes, which cause important changes in the behaviour of gold clusters.

The 'alcohol reduction process' is widely applicable to the preparation of colloidal precious metals stabilized by organic polymers such as poly (vinyl pyrolidone) (PVP), poly (Vinyl alcohol) (PVA) and poly (Vinyl methyel ether). Alcohols containing α–hydrogen atms are oxidized to the corresponding carbonyl compound during the salt reaction. Reduction of $RuCl_3$ in a liquid polyol of 1nm range, hydrogen is used as an efficient reducing agent for the preparation of electrostatically stabilized metal sols and of polymer stabilized hydrosols of Pd, Pt, Rh and Ir. Hydrogen reduction pathway gives Moiseev's giant Pd clusters, Finke's poly oxoanion and tetrabutyl ammonium stabilized transition clusters. Moiseev's giant cluster has the formula $Pd_{\sim 561}L_{\sim 60}$ (OAC) = 180 where L = phenanthroline or bipyridine.

Using 'CO', formic acid or sodium formate, HCHO, C_6H_5CHO as reductants, colloidal Pt in water is obtained. Tetrakis (hydroxyl methyl) phosphonium chloride (THPC) as a reducing agent allows the size and morphology of selective synthesis of Ag, Cu, Pt and Au nanoparticles from their metal salt. The reductant $[BEt_3H^-]$ is combined with the stabilizing agent (e.g NR_4^+). The surface active NR_4^+ salts are formed immediately at the reduction centre at high local concentration and prevent particle aggregation. Tri alkyl boron is recovered unchanged from the reaction and there are no borides contaminating the products. The chain length of the alkyl group in the tetra alkyl ammonium plays a critical role in the stabilization of various metal colloids.

$$MX_v + NR_4\,(BEt_3H) \rightarrow M_{colloid} + vNR_4X + vBEt_3 + v/2\,H_2\uparrow$$

where M = metals of groups 6–11; X = Cl^-, Br^-; v = 1,2,3; R = alkyl, C_6–C_{20}. The NR_4^+ stabilized metal 'raw' colloids as synthesized typically contain 6–12 wt% of metal. "Purified" transition metal colloids containing 70–85 wt% metal are obtained by workup with C_2H_5OH or ether and subsequent reprecipitation by a solvent of different polarity. When NR_4X is coupled to the metal salt prior to the reduction of step the preprecipitation $[NR_4^+\ Et_3H^-]$ can be avoided. Transition metal nanoparticles stabilized by $NR_4^+X^-$ is also obtained from NR_4X transition metal double salts. A number of conventional reducing agents may be applied since the local concentration of the protecting group is sufficiently high to give

$$(NR)_\omega MX_vY_\omega + vRed \rightarrow M_{Colloid} + vRedx + \omega\,NR_4Y$$

where M = metals; Red = H_2; HCOOH, K,Zn,LiH, $LiBEt_3H$, $NaBEt_3$, $KBEt_3$; X,Y = Cl, Br; v, ω = 1–3; R = alkyl; C_6 – C_{12}. Isolable metal colloids of the zerovalent early transition metals which are stabilized only with THF is prepared via the $[BEt_3H^-]$ reduction of the preformed THF adducts of $TiBr_4$, $ZrBr_4$, VBr_3, $NbCl_4$, and $MnBr_2$.

$$x[TiBr_4.2THF + x.4K(BEt_3H)] \xrightarrow[(THF,\ 2h,\ 20°C)]{} [Ti.O.5THF]_x$$
$$+ x.4BEt_3 + x.4KBr\downarrow + x.4H_2\uparrow$$

[Ti.O.5THF] is found to have Ti_{13} clusters in the zerovalent state stabilized by 6 intact THF molecules. **Table 1** summarizes the THF stabilized organosols of some transition metals.

TABLE 1. THF stabilized organosols of some transition metals.

Starting material	Reducing agent	Product	Duration (hour)	Metal content (%)
$TiBr_4$.2THF	$K[BEt_3H]$	[Ti.O.5THF]	6	43.5 (~ 0.8 nm)
$ZrBr_4$.2THF	$K[BEt_3H]$	[Zr.O.4THF]	6	42
VBr_3.3THF	$K[BEt_3H]$	[V.O.3THF]	2	51
$NbCl_4$.2THF	$K[BEt_3H]$	[Nb.O.3THF]	4	48
$MnBr_2$.2THF	$K[BEt_3H]$	[Mn.O.3THF]	3*	70 (~ 1 – 2.5 nm)

*50°C

1 to 2.5 nM size particles of [Mn.0.3THF] are also prepared. Tetra hydro thiophene is (THF) also used to synthesis Mn, Pd and Pt nanoparticles. **Table 2** gives an overview of $[BEt_3H^-]$ method.

TABLE 2. Overview of $[BEt_3H^-]$ method.

Stabilizer	Nanometals
THF	Ti, V, Zr, Nb, Mn
NR_4^+	Fe, Co, Ni, Cu, Ru, Rh, Pd, Ag, Re, Os, Ir, Pt, Au

$[BEt_3H^-]$ method has the advantages :

1. It is applicable to salts of metals in groups 4–11 in the periodic Table.
2. It yields extraordinary stable metal colloids that are easy to isolate as dry powders.
3. The particles size distribution is nearly mono disperse.
4. Bimetallic colloids are easily accessible by coreduction of different metal salts.
5. Suitable for scaling up and multigram production.

One of the drawbacks of this method is that the particles size of the resulting sols cannot be varied by altering the reaction conditions.Using betaines instead of NR_4^+ salts as the protecting group, highly water soluble hydrosols particularly those of zero valent precious metals are made.

Another method for the size and morphology selective preparation of metal colloids is using tetra alkyl ammonium carboxylates of the type $NR_4^+ R'CO_2^-$ [R-octyl, R′-alkyl, aryl, H] both as the reducing agent and the stabilizer.

$$M^+ + R_4N^+R'CO_2^- \xrightarrow[50-90°C]{} M°(R_4N^+R'CO_2)_x + CO_2 + R' - R$$

where, R-octyl, R′-alkyl, aryl and H. The electronic nature of the R′ group in the carboxylate group decides the size. Electron donors produce small nanoclusters with electron withdrawing substituents R′ yields larger particles. For example Pd particles of 2.2 nm size were found when Pd $(NO_3)_2$ is treated with an excess of tetra (n-octyl) ammonium carboxylate bearing $R'=(CH_3)_3\ CCOO^-$ as the substituent. The particle size is found to be 5.4nm with $R'=Cl_2\ CHCOO^-$ (an electron withdrawing substituent). Bimetallic colloids of the following are obtained with tetra (n-octyl) ammonium formates as the reductant: Pd/Pt (2.2nm), Pd/Sn(4.4 nm) Pd/Au (3.3nm), Pd/Cu(2.2nm). The shape of the particles also are found to depend on the reductant: with tetra (n-octyl) ammonium glycolate reduction of Pd $(No_3)_2$ trigonal particles of Pd colloid is observed **(Table 3)**.

Organo aluminium compounds can be used for the 'reductive stabilization' of mono and bimetallic nanoparticles.

$$MX_n + AlR_3 \xrightarrow[\text{Toluene}]{} [\text{Nanoparticle}] + [R_2Al\,\text{acac}]$$

where, acac = acetyl acetone.

Colloids of zerovalent elements of groups 6–11 of the Periodic Table (and also of tin) may be Prepared, in the form of stable, isolable organosols. The analytical data suggest that a layer of condensed organoaluminium species protects the transition metal core against aggregation. The exact nature of the 'backbone' of the colloidal organoaluminium protecting agent is not established.

TABLE 3. Mono and bimetallic nanocolloids in toluene via the organo-aluminium route.

Salt	Reducer	Particle size
Ni $(acac)_2$	Al $(i\text{-}but)_3$	2–4
Fe $(acac)_2$	Al $(me)_3$	—
$RhCl_3$	Al $(oct)_3$	2–3
Ag-decanoate	Al $(oct)_3$	8–12
Pt $(acac)_2$	Al $(me)_3$	2–5
$PtCl_2$	Al $(me)_3$	2.0
Pd $(acac)_2$	Al $(et)_3$	3.2
Pt $(acac)_2$		
Pt $(acac)_2$		
Ru $(acac)_2$	Al $(me)_3$	1.3
Pt $(acac)_2$		
$SnCl_2$	$Al(me)_3$	—

Quantitative protonolysis detected the presence of unreacted organo aluminum group (e.g. $Al–CH_3$, $Al–C_2H_5$) from the starting material. These active Al–C bonds in the colloidal protecting shell can react with inorganic surfaces bearing OH^- which opens new ways for the heterogeneous catalyst preparation. The particle size of the metal core is not altered during the modification process.

Electrochemical Synthesis

This is a versatile method of preparation of nanostructured mono and bimetallic colloids. The overall process involves 6 elementary steps:

1. Anodic dissolution of metal.
2. Migration of metal ions to the cathode
3. Reductive formation of zerovalent atoms at the cathode.
4. Formation of metal particles by nucleation and growth.
5. Arrest of the growth process and stabilization of the particles by colloidal protecting agents. Example tetra alkyl ammonium ions .
6. Precipitation of the nanostructured colloids.

Overall reaction :

Anode: $M \rightarrow M^{n+} + ne$

Cathode: $M^{n+} + ne + \text{stabilizer} \rightarrow M_{coll/stabilizer}$

Sum: $M + \text{Stabilizer} \rightarrow M_{coll/stabilizer}$

Electrochemical method offers, products are easily isolable; purity of the product; size selectivity; current density variation yielding smaller particles. Using Pd as the anode in a cell gives Pd particles of 6nm when $(C_8H_{17})_4 N^+Br^+$ is used as a stabilizer. NR_4^+ stabilized nickel particles yield smaller particles. This method is used to prepare monometallic organosols and hydrosols e.g. of Pd, Ni, Co, Fe, Ti, Ag and Au on a scale of several hundred milligrams. Pd particles (8–10nm) are obtained by propylene carbonate stabilizer. To get bimetallic (Pd/Ni, Fe/Co, Fe/Ni) nanocolloids, two anodes are used in a single cell. In the case of Pt, Rh, Ru and Mo which are anodically less readily soluble, the corresponding metal salts are reduced at the cathode **(Table 4)** Tetra alkyl ammonium acetate is used both as the supporting electrolyte and the stabilizer in a Kolbe electrolysis at the anode.

Cathode : $Pt^{2+} + 2e \rightarrow Pt$

Anode : $2CH_3COO \rightarrow 2CH_3CO_2 + 2e$

By modifying the electrochemical method, the synthesis of layered bimetallic nanocolloids e.g. (Pt/Pd) is achieved. A performed $(oct)_4$ NBr stabilized Pt colloid core (–3.8nm) is electrolysed in 0.1M $(oct)_4$ NBr/THF solution with Pd as the sacrifical anode. The performed Pt core may be regarded as a 'living metal polymer' on which the Pd atoms are deposited to give "onion type" bimetallic nanoparticles (5nm).

TABLE 4. Electrochemically prepared metallic/bimetallic colloids.

Salt	Size (nm)	Remarks
$PtCl_2$	2.5	c.d. 5mA/cm^2
$PtCl_2$	5.0	c.d. 0.05 mA/m^2
$RhCl_3.xH_2O$	2.5	
$RuCl_3.xH_2O$	3.5	
$OSCl_3$	2.0	

Pd $(OAc)_2$	2.5	
MO_2 $(OAc)_4$	5.0	
$PtCl_2$ + $RuCl_3.xH_2O$	2.5	
$PtCl_2$	3.0	PtSn alloy; when Sn anode is used
Pd $(OAc)_2$	2.5	CuPd alloy; when Cu anode is used
$PtCl_3$	3.5	PdPt alloy; when Pd anode is used

Transition Metal Complex

Under the action of heat, light or ultrasound low valency organometllic complexes and several organic derivatives of transition metals are decomposed to yield short lived nucleation particles of zerovalent metals in solution. Thermolysis leads to the rapid decomposition of cobalt carbonyls to give colloidal cobalt in organic solutions. Thermolysis of labile precious metals salts in the absence of stabilizers yields colloidal Pd, Pt and bimetallic. In the presence of stabilizing polymers such as PVP yields are good. Heating in a simple household microwave oven is used to prepare nanosize metal particles and colloids. The electromagnetic waves heat the substrate uniformly leading to more homogeneous nucleation and a shorter aggregation. Fe, MO_2C, Ni, Pd and silver nanoparticles in various stabilizing environments are made using sonochemical decomposition. By the controlled chemical decomposition of zerovalent transition metal complexes on the addition of CO or H_2 in the presence of appropriate stabilizers, isolable yields of colloidal product in multigram amounts are prepared. Low valency transition metal olefin complexes as a very clean source for the preparation of nanostructured mono and bimetallic methods are used for the nanoparticle colloids preparation.

The radiolytic synthesis of mixed Au (III)/Pd (II) solutions are used. At low dose rates a bilayer cluster with an Au/core/Pd shell predominates due to inter metal electron transfer form Pd atoms to gold ions, resulting first in the reduction of the latter to form the core of the particle and then a possible intermetal electron transfer, genuine alloyed clusters are formed. To control particle size distribution "size selective precipitation" is used. Using salt reduction method, size selective colloid synthesis is possible.

2.2 SYNTHESIS OF NANOCLUSTERS

Laser Induced Evaporation

A high intensity laser beam is incident on a metal rod causing evaporation of atoms from the surface of the mental. The atoms are then swept away from a burst of He and passed through an orifice into a vacuum where the expansion of the gas causes cooling and formation of clusters of the metal atoms. **Figure 2.1** presents the laser induced evaporation process. Under high pressure helium, the material (rod) is vapourized by laser beam and the particles enter the chemical reaction chamber and identified by mass spectrometer. This method is applicable to high melting point elements like C, Si and transition metals.

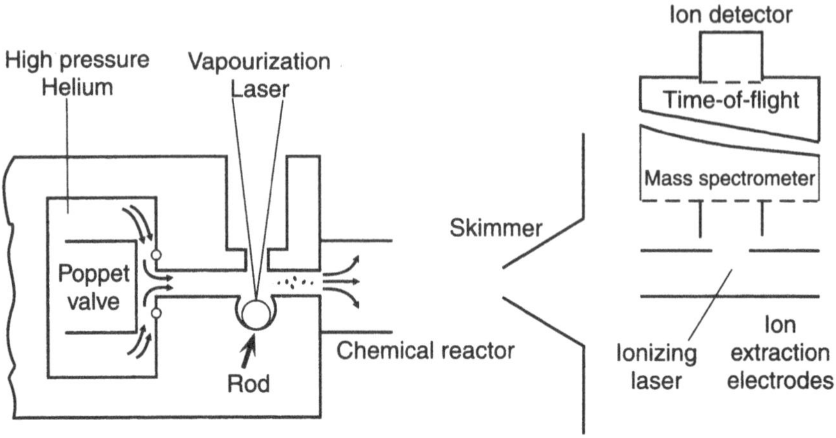

Fig. 2.1 Laser induced evaporation.

Pulsed Laser Methods

Pulsed lasers have been used in the synthesis of nanoparticles of Ag. $AgNO_3$ solution and a strong reducing agent are flowed through a blend like device. In the blender, there is a solid disk, which rotates in the solution. The solid disk is subjected to pulses from a laser beam creating hot spots on the surface of the disk. $AgNO_3$ and the reducing agent react at these hot spots resulting in the formation of small Ag particles which can be separated from the solution using a centrifuge. The size of the particle is controlled by the energy of the laser and the rotation speed of the disk. This method is capable of high rate of production of 2–3 g/min.

Laser Ablation

A high powered pulsed laser is utilized to evaporate materials from the surface of the source material kept in an inert gas chamber. Energy of the incident laser beam has to be selected for individual material clusters of semiconductors and metals have been successfully generated by this method. **Table 5** summarises metals and semiconductor clusters prepared by laser ablation method.

Chemical Methods

Several types of reducing can be used to produce nanoparticles such as NaB Et_3H, $LiBEt_3H$ and $NaBH_4$ where Et denotes the ethyl ($-C_2H_5$) radical. For example, nanoparticles of Molybdenum (Mo) can be reduced in toluene solution with NaB Et_3H at 25°C providing a high yield of Mo nanoparticles having dimensious of 1–5 nm.

$$MoCl_3 + 3NaBEt_3H \leftrightharpoons Mo + 3NaCl + 3BEt_3 + 11/2H_2\uparrow$$

TABLE 5. Metals and semiconductor clusters prepared by laser ablation method.

Cluster	Size (nm)
Au, pd	1–10
Er	10–70
Fe	5–40
CuCl	17–60
CdS	20–250

Nanoparticles of aluminium were made by decomposing $Me_2EtNAlH_3$ in toluene and heating the solution to 105°C for 2 hours ($Me = CH_3$). Titanium isopropoxide is added to the solution. The titanium acts as a catalyst for the reaction. The choice of catalyst determines the size of the particle produced. For example 80nm particles have been made using titanium. A surfactant like oleic acid can be added to the solution to coat the particles and prevent aggregation. **Figure 2.2** summarizes the

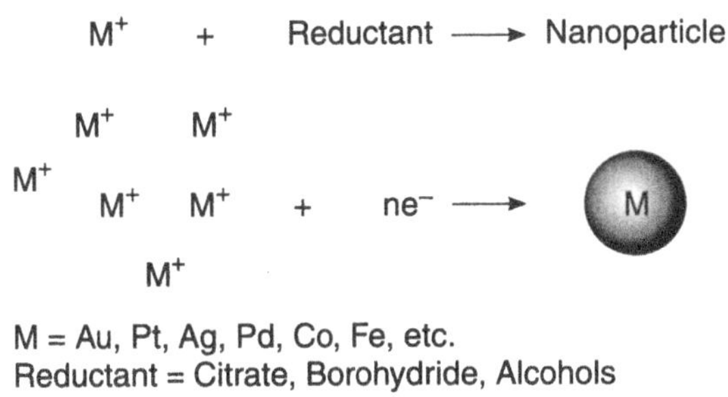

Fig. 2.2 Metallic nanoparticle synthesis.

metallic nanoparticle synthesis, Transition metal ions get reduced by citrate or borohydride or alcohols and form nanoparticles. These particles are to be prevented from growing further. Bimetallic or alloy particles are prepared by using two different metallic ions. Reductants maybe the same or different. Synthesis of bimetallic nanoparticle is shown as **Figure 2.3:**

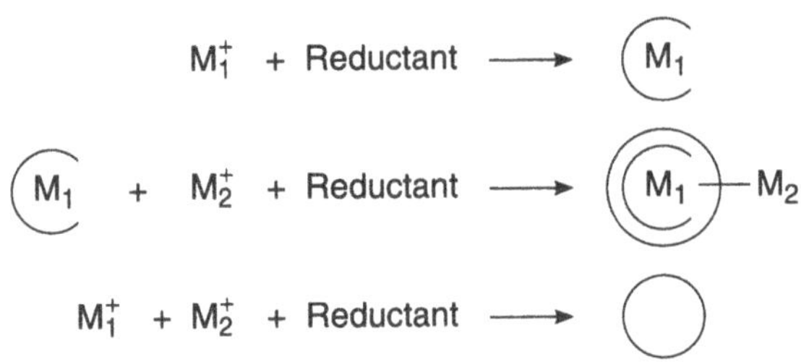

Fig. 2.3 Bi-metallic nanoparticle synthesis.

Many approaches are in vague like the use of colloids, micelles, polymers, glasses and crystalline hosts. Major problems with colloidal solutions are the irreproducibility in preparation, their stability and the difficulty in establishing the definite identity of the semiconductors. By incorporation of small semiconductor clusters into solid matrices such as polymers and glasses, the problems associated with the colloidal solutions are solved. The surfaces of the clusters are capped with organic or inorganic groups such that cluster is stable against agglomeration. Polyphosphate and thiols are the most commonly used capping agents. A molecular fragment of sphalerite CdS containing $Cd_{10}S_4$ core was capped by 16 thio phenolate groups. Cluster surfaces can be passivated by using hydroxide ions, amines, NH_3 and ZnS. On Thiophenolate capped CdS clusters, the size can be increased by adding extra sulphide to the solution (an example of inorganic living polymerization). The pyramidal shape cluster of $[Cd_{10} S_4 (SPh)_{16}]^{4-}$ with 30 Cd and S atoms can be made to a larger cluster $[Cd_{20} S_{13} (SPh)_{22}]^{8-}$. [55 Cd and S atoms]. This is known as cluster fusion. **Table 6** summarises a few clusters prepared by chemical methods.

TABLE 6. Nanoclusters prepared by chemical methods.

Material	Method	Cluster size (nm)
$MnFe_2O_4$	Coprecipitation	5–25
HgSe	Colloidal	2–10
PbI_2	Colloidal	1.2–3.10
Ag	Colloidal	4
ZrO_2, TiO_2	Gel precipitation Sol-gel	5–20
Al_2O_3–ZrO_2	Chemical polymerization and precipitation Sol-gel	20–80
$PbTiO_3$	Glass ceramics	12
Nd–Fe–Zr–B	Pyrolysis of	20–40
Fe–Co	organometallics	10–50

Thermolysis

It is decomposing solids containing metal cations, molecular anions or metal organic compounds. For example small lithium particles can be made by decomposing lithium azide, LiN_3. The material is placed in an evacuated quartz tube and heated to 400°C. At about 370°C, the LiN_3 decomposes releasing N_2 gas. Lithium atoms coalesce to form small colloidal metal particles. Particles <5nm can be prepared in this method. Passivation is achieved by introducing an appropriate gas. The presence of these nanoparticles can be detected by electron paramagnetic resonance (EPR). It is possible to estimate the size of the particles from the g-factor shift and line width of the spectra.

2.3 SYNTHESIS OF POLYMER SUPPORTED CLUSTERS

Polymer supported metal clusters can be prepared by a variety of methods. In a method a monomer unit is polymerized using an initiator. Micelles are formed from emulsifier. The resultant polymer unit is stabilized. (**Figure 2.4**). Generally there are two basic methods. One involving the in situ synthesis of the nanoparticles in presence of either the polymer dissolved in an appropriate solvent or a simple mixing between nanoclusters and the polymer with the help of sdvents. Extensive reorganization during these processes, can adversely affect the order and distribution of clusters. All these approaches basically involve 3 common steps as shown below :

(a) Preparation of monomer/polymer compatible with the nature of capping of nanoparticles.

(b) Synthesis of nanoparticles using suitable precursors, and

(c) Organization of nanoparticles in a polymer matrix using different synthetic approaches.

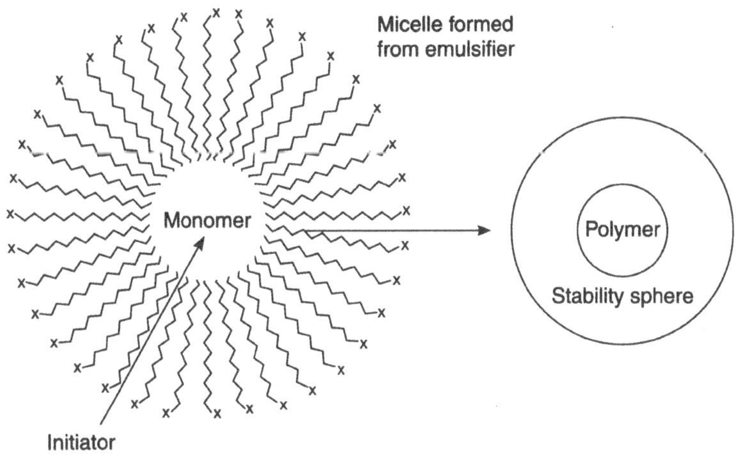

Fig. 2.4 Polymer nanoparticle synthesis.

In most of the cases, the metal is introduced in the preformed polymer bound functionalities with suitable metal precursors. Metal precursors can subsequently and conveniently be reduced to form either polymer supported nanoparticles or nanoparticles with the polymer. This way of nano composite preparation is usually accompanied by the reduction of the polymer bound metal precursors in a compatible solvent and common reducing agents like $NaBH_4$, hydrazine, alchol, HCHO and H_2 etc., for the case of metal clusters.

In some case of polymers, nanoparticles can be assembled only on the surface because even if after separate procedure, both are mixed in various proportions, clusters inevitably segregate on the surface. This can be contrasted with certain other type of polymers where nanoparticles get bound inside the matrix of the polymer so that various combinations can be prepared to suit the type of applications. For example Fe–Pt nanoparticles on the surface of active polymers like poly – (vinylpyrrolidene) and poly–(ethyleneamine) has been developed for magnetic reversal behaviour.

Selection of Polymers

The basis of formation of metal-polymer nanoparticles is the interaction of polymer functional group with metal nanoparticles. A variety of polymers can be selected depending on the nature and amount of functional groups structurally compatible with the monolayer capping of the nanoclusters, if functionalized clusters are used. Polymers with functional groups like – SH – SO_3H – COOH – NH_2 etc. can directly link with Au, Ag, Cu, Pt and other similar nanoclusters. A wide variety of structures can be selected for different types of monomer (co-monomer) ratios and even block copolymers can be used for obtaining certain specific periodic functions. For example, coblock polymers having polystyrene and poly (2-vinyl) pyridine or polystyrene and polypyrole with controlled hydrophobic/hydrophilic excellent means to obtain kinetically controlled dispersion where metal/semiconductor clusters show

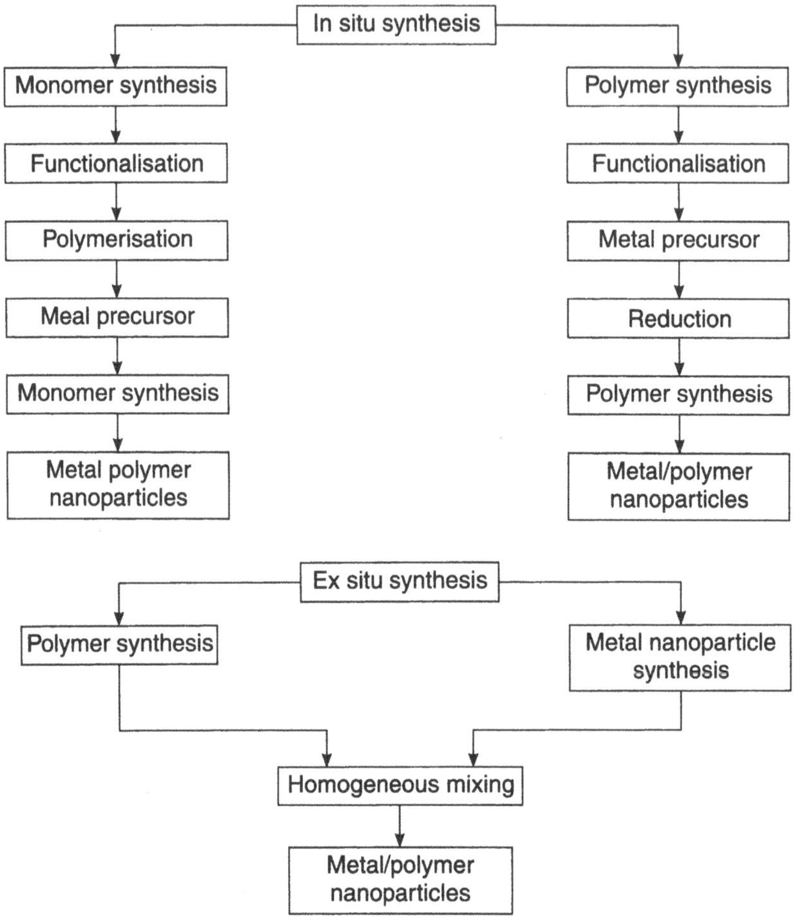

Fig. 2.5 Flow sheet diagram of different preparation/organization strategies of nanoparticles on polymer surfaces.

specific affinity towards one of the blocks. Consequently nanoclusters can be arranged on one of these regions and the periodic formation of such hydrophobic / hydrophilic blocks at optical wavelength scales can have important implications in design of photonic bandgap materials. Block copolymers can be synthesized with a very narrow molecular weight distribution and with a variety of macromolecular architectures.

Another important type of polymer useful for cluster organizations can be selected on the basis of electronic conductivity. The selection of π conjugated polymers having strong electron donating properties even allows in situ formation of nanoparticles by the reduction of metal ions via electron transfer from π conjugated polymer to metal ions. The electronic properties of these nano composites can be controlled during the synthesis stage itself since these polymers strongly influence the characteristics of the nanoparticles and provide a potentially efficient route for shultling of electronic charge between the nanoclusters. Design of several three dimensional polymeric sturctrures with multiple internal and external functional groups to act as host for a range of nanoclusters is possible. Several interesting types of dendrimers structures have been found to retain clusters selectively depending upon the size, shape and distribution. This way of designing dendrimer-nanocluster composites have applications in bioelectronics. Table 7 gives a summary of the different nanoparticles capped with polymers having appropriate functional group and their specific features.

TABLE 7. Summary of the different nanoparticles capped with polymers having appropriate functional groups.

Metal clusters	Polymer support	Size and features
	(isopropyl)	3-9nm
Gold	Acrylamide	6-7nm, crystalline
	Polystyrene	5-6nm, Semicrystalline
	Poly (amido amine)	
	Polymethyl methacrylate	50-70nm, Semicrystalline
	Polyamidoamine	
	Polystyrene-b-poly	2-7nm, Crystalline
Platinum	(2-Vinyl pyridine)	
	Polyethylene glycol	
	Polystyrene	1-2nm, Crystalline
Silver	Polystyrene/ methacrylate	100nm, Core shell
Cobalt	Polystyrene	2-3nm, Semicrystalline
Palladium	Poly (4-Vinyl) Pyridine	3-7nm
Fe, Co	Polyimine	10-30nm
Manganese	Norbornene related Polymer	10-15nm
CdSe	Polythiophene	10-15nm

2.4 SYNTHESIS OF NANOTUBES

Nanotubes of dichalogenides such as MOS_2, $MoSe_2$ and WS_2 are also obtained by employing processes far from equilibrium, such as are discharge and laser ablation. MOS_2 and WS_2 nanotubes are conveniently prepared by starting with the stable oxides MoO_3 and WO_3. The oxides are first heated at high temperatures in a reducing atmosphere and then reacted with H_2S. H_2Se is used to get selenides. Trisulphides are directly decomposed to obtain the disulphide nanotubes. Diselenides nanotubes are obtained from the metal triselenides. The triselenide route is found to provide a general route for the synthesis of the nanotubes of many metal disulphides such NbS_2 and HfS_2. In the case of Mo and W dichalcogenides, it is possible to use the decomposition of the precursor ammonium salt such as $(NH_4)_2\ MX_4$ (x = S, Se; M = Mo, W) as a means of preparing the nanotubes. Other methods employed for the synthesis of dichalcogenide nanotubes include hydrothermal methods where the organic amine is taken as one of the components of in the reaction mixture. The hydrothermal route is used for synthesizing nanotubes and related structures of a variety of other inorganic materials as well. Thus nanotubes of several metal oxides (e.g. SiO_2, V_2O_5, ZnO) are produced hydrothermally. Nanotubes of oxides such as V_2O_5 are also conveniently prepared from a suitable metal oxide precursor in the presence of an organic amine or a surfactant for example CdSe and CdS. The metal oxide reacts with the sulphiding / selenidizing agent in the presence of a surfactant such as Triton X.

Sol-gel chemistry is widely used in the synthesis of metal oxide nanotubes (SiO_2 and TiO_2). Oxide gels in the presence of surfactants or suitable templates form nanotubes. By coating nanotubes (CNT) with oxide gels and then burning off the carbon, one obtains nanotubes and nanowires of a variety of metal oxides including ZnO, SiO_2 & MoO_3. Sol-gel synthesis of oxide nanotubes are possible in the pores of alumina membrane. MoS_2 nanotubes are also prepared by the decomposition of a precursor in the pores of an alumina membrane.

BN nanotubes are obtained by striking an electric arc between HfB_2 electrodes in N_2 atmosphere. BCN and BC nanotubes are obtained by arcing between B/C electrodes in an appropriate atmosphere. BN nanotubes are made by using different precursor molecules containing B and N. Decomposition of borazine in the presence of transition metal nanoparticles and the decomposition of the 1:2 melamine – boric acid addition compound yield BN nanotubes. Reaction of H_3BO_3 or B_2O_3 with N_2 or NH_3 at high temperature in the presence of activated carbon, carbon nanotubes are made. Catalytic metal particles are employed to synthesize BN nanotubes. Single crystal GaN nanotubes with inner diameters of 30–200nm and wall thickness of 5–50nm by employing "epitaxial casting" approach are made. For this hexagonal ZnO nanowires as templates for this epitaxial overgrowth of thin GaN layers in a chemical vapour deposition system are employed. Trimethyl gallium and NH_3 as precursors with Argon or N_2 gas as carrier gas and maintained the decomposition temperature at 600–700°C are used. The ZnO nanowires templates were subsequently removed by thermal reduction and evaporation resulting in ordered arrays of GaN nanotubes on the substrates.

Organic templates (Organosols) are comprised of an organic liquid and low concentrations of (relatively low) low molecular weight molecules. These are prepared by heating a mixture of a gelator and solvent until the solid dissolves upon cooling the solution (Sol) thickens to form a gel. The gelation of a number of compounds such as tetra ethyl ortho silicate (TEOS) along with a cholesterol based gelator under acidic pH conditions followed by poly condensation, exhibits a network of fibers with diameter ranging from 50–200nm. Subsequent drying and calcinations steps resulted in silica tubes without the presence of the original organic template. **Table 8** lists some of the important inorganic nanotubes syntheses and characterized by either SEM or TEM or HREM.

CVD techniques are used to synthesis BN nanotubes and nanowires. **Figure 2.6** shows a typical chemical vapour deposition furnace where reacting gases are introduced on a quartz tube, water on which the particle has to be deposited are kept

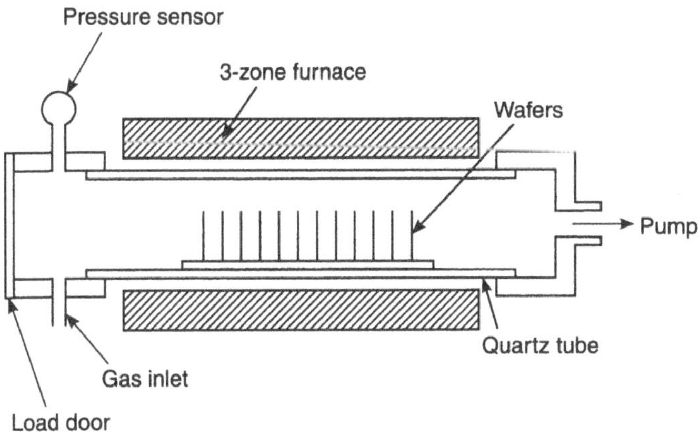

Fig. 2.6 Chemical vapour deposition—typical CVD furnace.

in a 3 zone furnace. The methods examined include heating H_3BO_3 with activated carbon, multiwalled carbon nanotubes, catalytic iron particles or a mixture of activated carbon and iron particles in the presence of NH_3. With activated carbon, BN nanowires are obtained as the primary product. However with multiwalled carbon tubes, high yields of pure BN nanotubes are obtained as the major product. BN nanotubes with different structures were obtained on heating H_3BO_3 and iron particles in the presence of NH_3. Aligned BN nanotubes are obtained when aligned multiwalled nanotubes are used as the templates.

Needle–shaped TiO_2 (anatase) nanotubes can be precipitated from a gel containing a mixture of SiO_2 and TiO_2. A mixture of titanium isopropoxide and TEOS, was hydrolysed and gelled in an cubator and the gel further heated to 600°C resulting in the precipitation of fine TiO_2 crystals. This was further treated with NaOH at 100°C for 2 hours to yield the TiO_2 nanotubular phase. An amorphous SiO_2 related phase present in the product was removed by chemical treatment. The nanotubes formed by this method had a diameter of ~ 8nm and lengths upto 100nm. Much smaller TiO_2 nanotubes was also synthesized. TiO_2 with anatase or rutile structure was treated with NaOH and subsequently with HCl. The resulting TiO_2 nanotubes are 50–200nm long and their diameter is ~10nm. The HREM image of such a TiO_2 nanotubes shows that the tubes have an inner core and walls. The presence of lattice fringes indicates that the crystalline structure of TiO_2 nanotubes. Parallel fringes in the walls correspond to a distance of ~7nm which are detected by X-ray and electron diffraction.

Nanotubes of II–VI semiconductor compounds such as CdS and CdSe have been obtained by a soft chemical route involving surfactant assisted synthesis for CdS nanotubes. The metal oxide was reacted with the sulfidizing reagent in the presence of a surfactant such as Triton 100x in a basic medium. To obtain nanotubes of CdSe, similar procedure is used except that NaHSe was used in the place of thioacetamide as selenidizing reagent in the presence of a surfactant reagent such as Triton 100x. Both the CdSe and CdS nanotubes seem to be polycrystalline formed by aggregates of nanoparticels. The nanotubes of CdSe though extended in one direction show quantum confinement and the absorption band is blue shifted to 550nm from 650nm in the bulk sample.

By hydrothermal method, alkyl ammonium intercalated VO_x can be synthesized. The Vanadium alkoxide precursor was hydrolysed in the presence of hexa decylamine and the hydrolysis product (lamellar structured composite of the surfactant and the Vanadium oxide) yielded VO_x nanotubes along with the intercalated amine under hydrothermal conditions. In this Vanadium oxide nanotube, Vanadium is in the mixed Valent state there by rendering it redox active. The template could not be removed by calcinations as the structural stability was lost above 250°C. Under mildly acidic conditions, the surfactant can be extracted. The alkylamine intercalated in the inter tubular space could be exchanged with other alkylamines of varying chain lengths as well as α, ω-diamines. The distance between the layers in the VO_x nanotubes can be controlled by the length of the $-CH_2$ chain in the amine template.

Most of the VO_x nanotubes got by the hydrothermal method are open ended. Very few closed tubes had flat or pointed conical tips. Cross sectional TEM images of the nanotubular phases show that instead of concentric cylinders (i.e. layers that fold and close with themselves) the tubes close with single or double layer scrolls providing a ser pentine – like morphology. The scrolls are usually seen in the images as circles. Non symmetric fringe patterns in the tube walls exemplify that most of the nanotubes are not rotationally symmetric and carry depressions and holes in the walls. Diamine intercalated VO_x nanotubes are multilayer scrolls with narrow cores and thick walls composed of packs of several vanadium oxide layers. Many of the nanotubes formed by layered materials show various types of defects. They exhibit unusual tip structures. The tips are not always spherical in these nanotubes.

2.5 SYNTHESIS OF CARBON NANOTUBES

Techniques have been developed to produce nanotubes in sizable quantities including arc discharge, laser ablation high pressure Carbon Monoxide and chemical vapour deposition. All of these processes take place in vacuum with process gases. Nanotubes can be synthesized using a carbon arc. A potential of 20–25V is applied across carbon electrodes of 5–20 μm diameter and separated by 1mm at 600 torr pressure of flowing He. Carbon atoms are ejected from the positive electrode and form nanotubes on the negative electrode. As the tubes form, the length of the positive electrode decreases. Carbon is deposited on the negative electrode. To produce SWNT, a small amount of cobalt, nickel or iron is incorporated as a catalyst in the central region of the positive electrode. If no catalysts are used, the tubes are nested or multiwalled. The carbon arc method can produce SWNT of 1–5nm with a length of 1μm.

In the laser ablation process, a pulsed laser vapourises a graphite target in a high temperature reactor while an inert gas is blowed into the chamber. The nanotubes develop on the cooler surfaces of the reactor as the vapourised carbon condenses. A water cooled surface may be included in the system to collect the nanotubes.

Decomposing a hydrocarbon gas such as CH_4 at 1100°C by chemical vapour deposition can produce carbon nanotubes. As the gas decomposes, carbon atoms are produced that then condense on a cooler substrate that may contain various catalysts such as iron. This method produced tubes with open ends which does not occur when other methods are used. This procedure allows continuous fabrication and scale up. The metal catalyst is necessary for the growth of SWNT, namely cobalt or nickel. The atoms of the metal catalyst attach to the dangling bonds at the open end of the tubes and the atoms "scoot" around the rim of the tube adsorbing as they arrive. This is known as "Scooter mechanism".

The diameters of the nanotubes that are to be grown are related to the size of the metal particles. This can be controlled by paltered (or masked) deposition of the metal annealing or plasma etching of a metal layer. The substrate is heated to approximately 700°C. To initiate the growth of nanotubes, two gases are bled into the reactor: a process gas (example NH_3, N_2, H_2 etc.,) and a carbon containing gas (example NH_3, N_2, H_2 etc.). Nanotubes grow at the sites of the metal catalyst, the carbon containing gas is broken apart at the surface of the catalyst particle and the carbon is transported to the edges of the particle where it forms the nanotubes. The catalyst particles generally stay at the tips of the growing nanotube during the growth process although in some cases they remain at the nanotube base depending on the adhesion between the catalyst particle and the substrate.

If a plasma is generated by the application of a strong electric field during the growth process [plasma enhanced CVD] then the nanotube growth will follow the direction of the electric field. By properly adjusting the geometry of the reactor it is possible to synthesise vertically aligned carbon nanotubes (i.e., perpendicular to the substrate) a morphology that has been interest in the field of electron emission from the nanotubes. Without plasma the resulting nanotubes are randomly oriented, resembling a bowl of spaghetti. CVD method offers an advantage to grow on a desired substrate.

Carbon nanotubes may be semiconducting or conducting. To separate them, IBM had developed a method. It was accomplished by depositing bundles of nanotubes on a silicon wafer. Metal electrodes were then deposited over the bundle. Using the silicon wafer as an electrode, a small bias voltage was applied that prevent the semiconducting from the conducting effectively making them insulators. A high voltage is then applied across the metal electrodes thereby sending a high current through the metallic tubes but not the insulating tubes. This causes the metallic tubes to vapourise leaving behind only the semiconducting tubes.

Fullerenes and carbon nanotubes are not necessarily products of high-tech laboratories; they are commonly formed in such mundane places as candle flames. But for well defined qualities, process controlled methods are used.

2.6 SYNTHESIS OF NANOWIRES

One of the aspects of 1D structures relates to crystallization is nucleation and growth. As the concentration of the building units (atoms, ions or molecules) of a solid becomes sufficiently high, they aggregate into small nuclei or clusters through homogeneous nucleation. These clusters serve as seeds for further growth to form larger clusters. Several chemical strategies are developed for 1D nanowires with different levels of control over the growth parameters. These include: (1) the use of the anisotropic crystallographic structure of the solid to facilitate 1D nanowire growth (2) the introduction of a solid-liquid interface to reduce the symmetry of a seed (3) the use of templates (with 1D morphologies) to direct the formation of nanowires (4) the use of capping agents to kinetically control the growth rates of various facets of a seed and (5) self assembly of OD nanostructures. They can be usefully categorized into (1) nanowire growth in the gas phase (2) solution based approaches to nanowires.

Vapour Phase Growth

Using simple evaporation technique in an appropriate atmosphere elemental or oxide nanowires are produced.

Vapour-Liquid-Solid (VLS) growth : The growth of a nanowire via a gas phase reduction involving a VLS process is known. An anisotropic crystal growth is promoted by the presence of a liquid alloy-solid interface. The growth of Ge nanowire using Au cluster as solvent at high temperatures is known. The Ge and Au will form a liquid alloy when the temperature is higher than the eutectic point (363°C). The liquid surface has a large accommodation coefficient and is a preferred deposition site for incoming Ge vapour. After the liquid alloy becomes super saturated with Ge, Ge nanowire growth occurs by precipitation at the solid-liquid interface. A real time observation of Ge nanowire growth conducted in an "in situ" high tempertature transition electron microscope revealed a sequence of TEM images which directly mirrors the VLS mechanism. The VLS method is used to produce 1–100μM diameter 1D structures (whiskers). By controlling the nucleation and growth, it is possible to produce semiconductor nanowhiskers (e.g., In As, GaAs) II–VI semiconductors, oxides. A laser ablation based VLS process is primarily to produce semiconductor nanowires. TEM studies showed that the product obtained after the VLS growth is wire like structures with remarkably uniform diameters of the order of 10nm with lengths upto >1μM. VLS method is used to synthesis highly pure ultra long and uniform sized semiconductor nanowires in bulk quantities by employing laser ablation and thermal evaporation of semiconductor powders mixed with metal or oxide catalysts. The solid target used for laser ablation is made of pure Si powder mixed with metals (Fe, Ni or Co) and the temperature around the solid target is maintained in the range of 1200–1400°C. The temperature of the area around the substrate on which the nanowire grew is between 900–1100°C. TEM pictures show that Si nanowires obtained by this method are extremely long and highly curved with a typical diameter in the range of 20–80nm. Each wire consists of an outer layer of Si oxide and a crystalline Si core. A high density of defects such as stacking faults and micro twins is observed in the crystalline Si core. The axis of the nanowires are generally along the <112> direction and the {111} surfaces of Si crystalline cores are parallel to the axis of the nanowire.

Oxide assisted growth : Another concept is oxide assisted nanowire growth. Using this, synthesis and optical characterization of GaAs nanowires have lengths upto tens of micrometers and diameter in the range of 10–120nm with an average of 60nm are possible. The nanowires have a thin oxide layer covering and a crystalline GaAs core with a <111> growth direction. The oxide assisted nanowire growth mechanism is applied to the production of Si nanowires. The growth of Si nanowires is greatly enhanced when SiO_2 containing Si powder targets are used.

Highly pure Si powder is used for synthesizing Si nanowires by laser ablation. A large quantity of Si nanowires is obtained by mixing 30–70% SiO_2 into the Si powder target. SiO_2 played a crucial role in enhancing the formation and growth of the Si nanowires. In the proposed oxide growth mechanism, the vapour phase of $Si_xO(0<x>1)$ generated by thermal evaporation or laser ablation is the key factor.

$$Si_xO_{(S)} \rightarrow Si_{x+1(S)} + SiO_{(S)} \ (x>1)$$

$$2SiO \rightarrow Si_{(S)} + \ Si$$

The Si nanowires synthesized by using different SiO_2 contents in the targets are similar in structure except that the outer-Si oxide surfaces of the nanowires synthesized from targets with high SiO_2 content are rough. The diameter of the nanowires measured from the TEM image ranges from 9 to 12nm. Most nanowires are Quasi-aligned. The intensity of the cubic Si(111) diffraction ring exhibits a strong texture feature of the Si crystals in the nanowires. This indicates that the crystals in the nanowires should have a similar orientation, i.e., the Si nanowires should have a similar growth direction.

Besides the VLS mechanism, the classical vapour-solid (VS) method for whiskers growth also merits attention. In this method, the vapour is first generated by evaporation, chemical reduction or gaseous reaction. The vapour is subsequently transported and condensed into a substrate. The VS method is used to prepare oxide, metal whiskers with micrometer diameters. Hence it is possible to synthesize the 1D nanostructures. Using the VS process one can control its nucleation and subsequent growth process. Synthesis of nanowires for oxides of Zn, Sn, In, Cd, Mg and Ga are used. Synthesis of crystalline GaP nanowires with a mean diameter of 40nm and length upto 300μm via sublimation of ball milled GaP powder. Simple sublimation of SiO_2 powder can give large quantities of Si nanowires. The themal sublimation of SiO powder produced SiO vapour, which under went a disproportional reaction is transported and deposited at 930°C to form nanowires containing a crystalline Si core and an amorphous SiO_2 sheath. The axis of the Si nanowires is approximately along the [211] direction. This method has the advantage over the laser assisted catalytic growth method because it can produce high purity Si nanowires with out any metal contamination. Si nanowires thus prepared are stable faceted. STM of these Si wires showed automatically resolved images with two types of nanowire surfaces which are due to hydrogen-terminated Si(111)-(1X1) and Si(001)-(1X1) surfaces corresponding to SiH_3 on Si(111) and SiH_2 on Si(001) respectively. Hydrogen terminated Si nanowire surfaces are more oxidation-resistant than similarly treated Si wafer surfaces. The STS measurements shows that the electronic energy gaps are found to increase with decreasing Si nanowires diameter from 1.4eV for 7nm to 3.5eV for 1.3nm.

Carbothermal Reactions

A variety of oxides, nitrides and elemental nanowires can be synthesized. Carbon (activated carbon or carbon nanotubes) in mixture with an oxide produces oxide or sub oxide vapour species which react with other reactants (O_2, N_2 or NH_3) to produce the desired nanowires. GaN nanowires are produced by heating a mixture of Ga_2O_3 and carbon in N_2 or NH_3. Silicon nanowires are made by heating SiO_2 with carbon in a suitable atmosphere.

The simplest method to obtain β-SiC nanowires involves heating silica gel with activated carbon at 1360°C in H_2 or NH_3. In the presence of catalytic iron particles at 1200°C it gives α-Si_3N_4 nanowires and Si_2N_2O nanowires at 1100°C. Si_3N_4 can also be obtained by heating MWNTS with silica gel at 1360°C in an atmosphere of NH_3. In the presence of catalytic iron particles this method yields Si_3N_4 nanowires in pure form. The formation of carbide follows two steps : initially carbon reduces the SiO_2 to the volatile suboxide of silicon and then the formation of carbon as follows :

$$SiO_2 + C \rightarrow SiO + CO$$

$$SiO + 2C \rightarrow SiC + CO$$

By using the correct carbon source with the silica gel under carbothermal conditions it is possible to get nitride or carbide nanowires. With the carbon nanotubes the reaction follows one step to produce $Si_3\ N_4$ nanowires.

$$3SiO_2 + 6C + 4NH_3 \rightarrow Si_3N_4 + 6H_2 + 6CO$$

The role of the catalytic iron particles in the above reactions is likely to be in facilitating the removal of oxygen from the silica. The iron oxide formed in such a reaction would be readily reduced back to metal particles in the reducing atmosphere. Different nanostructures of β-Ga_2O_3 are prepared by the reaction of gallium oxide with activated carbon and carbon nanotubes. The flow rate of the Ar gas determines the morphology of the final nanostructures, thin nanowires being favoured by a high flow rate of Ar where as at very low rates of Ar nanotubes of Ga_2O_3 are obtained with high yield. The reaction of Ga_2O_3 powder with activated carbon mainly gives rise to nanosheets and nanorods. The HREM image shows Ga_2O_3 nanowired as single crystalline with the growth direction perpendicular to the (102) planes. Gallium nitride nanowires are also prepared by using carbon nanotubes templates or catalytic Fe(Ni) metal particles. These GaN nanowires are single crystalline with the wurt size structure and have high aspect ratios with lengths in the μm range.

Nanowires of metal silicides are also prepared. Submonolayer amounts of Er deposited onto Si(001) react with the substrate to form epitaxial nanowires of crystalline $ErSi_2$. The $EiSi_2$ nanowires so deposited are < 1nm high, a few nm wide, close to a micron long, crystallographically aligned to Si <110> directions. Using a CVD method, core shell and core multishell nanowire hetero structures are prepared. By this method the nanowires are grown by gradually building up thin uniform shells around a nm sized cluster of gold atoms. The nanowires wrapped around a silicon oxide core. These are only 50nm in diameter containing a germanium core surrounded by a silicon shell.

Solution Phase Method

The synthetic strategy mainly involves: (1) Anisotropic growth dictated by the crystallographic structure of a solid material; (2) Use of templates; (3) Kinetically controlled by super saturation; and (4) Use of appropriate capping agents.

Many solid materials such as polysulphur nitride $(SN)_x$ grow into 1D nano structures and their habit is determined by the highly anisotropic bonding in the crystallographic structure. Se, Te and molybdenum chaliogenides are obtained as nanowires due to the anisotropic bonding, which makes the crystallization occur along the C-axis favoring the stronger covalent bonds over the relatively weak Vander Waals forces between the chains.

Molybdenum chaliogenides with the general formula $M_2 MO_6 X_6$ [M=Li, Na; X=Se, Te] contain hexagonally close packed linear chains of formula $MO_6 X_6$. When dissolved in a highly polar solvent such as DMF or DMSO, M-mehyl formamide they mainly exist as chains of ~2nm diameter. Some chains may aggregate into bundles of fibers with cross sections of ~1μm diameter and lengths upto ~1μm. A spherical colloidal suspension/dispersion of amorphous (a-) selenium with diameters of ~300nm is refluxed with selenius acid and hydrazine at elevated temperatures. After cooling, the suspension to room temperature a small amount of Se dissolved in th solution precipitates out as nanocrystals of trigonal Se (t-Se). During align of this dispersion in the dark, the a-Si dissolves slowly in the solution and subsequently crystallizes out slowly on at –Se seed. The intrinsic anisotropic nature of t-Se building blocks, that is extended, helical chains of Se atoms in the trigonal phase assemblies (crystallizes) ultimately into naowires of t-Se.

Template Based Synthesis

In this technique, the template simply serves as a scaffold against which other kinds of materials with similar morphologies are synthesized. In situ generated material is shaped into a nanostructure with its morphology complementary to that of the template. These templates could be nanoscale channels within mesoporous materials or porous alumina and poly carbonate membranes. They can be filled using : (1) A solution route or (2) A sol-gel technique or (3) An electrochemical route to generate 1D nanowires.

The produced nanowires can be released from the templates by selectively removing the host matrix. Unlike the polymer membranes fabricated by track itching, porous AAo membranes containing hexagonally packed 2D array of cylindrical pores with a uniform size are prepared using anodized aluminium foils in acidic medium. Many materials are fabricated into nanowire using porous anodic alumina membranes (AAM) in a templating process including various inorganic materials such as Au, Ag,

Pt, TiO_2, MnO_2, ZnO, SnO_2, electronically conducting polymers, poly pyrole and others. Besides alumina and polymer membranes, with their high surface areas and uniform pore sizes, mesoporous silica materials are used as templates for the synthesis of polymer and inorganic nanowires. Ag nanowires of uniform diameters of 5-6nm and large aspect ratios between 100 and 1000 are synthesized by thermal decomposition. Ge nanowires are synthesized within the meso channels of MCM-41.

Meso phase structures self assembled from surfactants provide another class of useful and versatile templates for generating 1D nanostructures in relatively large quantities. It is well known that at critical micellar concentration (CMC), Surfactant molecules spontaneously organize into rod shaped micelles. These anisotropic structures can be used immediately as soft templates to promote the formation of nanorods when coupled with appropriate chemical or electrochemical reaction. The surfactants needs to be selectively removed to collect the nanorods/nanowires as a relatively pure sample. Based on this principle, nanowires of CuS, CuSe, CdS, CdSe, ZnS, ZnSe are grown selectively by using surfactants such as Na-AOT or triton-x of known concentrations.

The nanowires themselves are used as templates to generate the nanowires of other materials. The template may be coated onto the nanowire forming coaxial nanocables or it may react with the nanowires forming a new material. In the physical (solution or sol-gel coating) approach, the surfaces of the nanowires can be directly coated with conformal sheaths made of a different material to form coaxial nanocables. Subsequent dissolution of the original nanowires can lead to nanotubes of the coated material. The sol-gel coating method is a generic route to syntheses coaxial nanocables that may contain electrically conductive metal cores and insulating sheaths. The thickness of the SiO_2 sheath could be controlled in the range of 2–100nm by varying the concentration of the precursor and the deposition time.

Single crystalline Ag_2Se (tetragonal) nanowires of diameters less than 40nm are synthesized through a novel topotactic reaction when an t-Se single crystalline nanowire templates react with $AgNO_3$ solutions at 25°C. The high resolution TEM image obtained from the edge of an individual nanowires reveals the complete conversion of Se nanowire into single crystalline tetragonal Ag_2Se nano wire. The fringe spacing of 0.25nm corresponding to the interplanar distance of [200] implying the growth direction of this nanowire is $<100>$. Beyond 40nm diameter the orthorhombic structure becomes more stable. In some other cases this technique yields products of a polycrystalline nature. Template directed synthesis of metal nanorods covered by carbon and other materials is known. Nanowires of Au, Pt, Pd and Ag are prepared by employing sealed tube reactions as well as solution methods. Incorporation of thin layers of metals in the inter tubular space of the SWNT bundles are known.

A metal (e.g. In, Sn, Bi) with a low melting point is used as a catalyst and the desired material is generated through the decomposition of organo metallic precursors. Usually single crystalline whiskers or filaments with dimensions of 10–150nm and length upto several microns are made. Highly crystalline InP fibers of diameter

10–100nm and length 50–1000nm are grown by the methanolysis of {tert, Bu_2 In [μ-$P(SiMe_3)_2]_2$} in aromatic solvent at 111-203°C. The key component of the synthesis is the decomposition of the organometallic precursor which proceeds through a sequence of isolated and fully characterized intermediates to yield a complex [tert-$Bu_2In(\mu\text{-}PH_2)]_3$; this complex subsequently undergoes alkane elimination to generate the building blocks $(InP)_n$ fragments. The fragments dissolved in a dispersion of droplets formed by molten indium and recrystallised as InP fibers.

Uniform diameters ranging from 40–50A° (Si), 50–300A° (Ge) with several lengths of several μm of silicon and germanium are grown by the supercritical fluid-liquid–solid method (SFLS) method. Solvent dispersed size mono disperse alkane thiol-capped gold (Au) nanocrystals to direct the Si nanowire growth with narrow wire diameters distributions. Sterically stabilized Au nanocrystals are dispersed in supercritical hexane with a silicon precursor, diphenyl silane at a temperature of 500°C and pressure of 270 bar. At this temperature, the diphenyl silane decomposes to Si atoms which dissolve into the sterically stabilized Au nanocrystals until reaching supersaturation, at which point they are expelled from the particle as a thin nanometer scale wire. This supercritical fluid medium provides the high temperatures necessary to promote Si crystallization.

Ag nanowires are produced by reducing $AgNO_3$ with ethylene glycol in the presence of polyvinyl pyrolidone (PVP). They key to the formation of a 1D nanostructure is the use of PVP as a polymeric capping reagent and the introduction of a seeding step. $AgNO_3$ reduces in the presence of the seed (Pt or Ag nanoparticles) to form Ag nanoparticles with bimodal size distribution produced via homogeneous and heterogeneous nucleation processes. Ag nanorods grow at the expense of the small Ag nanoparticles as directed by the capping reagent. In the presence of PVP, most silver particles can be confined and directed to grow into nanowires of uniform diameter. These have FCC structure with a mean diameter of ~40nm.

Solvo Thermal Synthesis

A solvent is mixed with certain metal precursors and possibly crystal growth regulating or templating agents such as amines. The solution mixture is placed in an autoclave kept at relatively high temperature and pressure to carry out the crystal growth and assembly process. Many semiconductor nanorods and wires are prepared using this method.

Growth control : A challenge in synthesis is to control dimensions as the physical and thermodynamical properties depend in size. Controlled synthesis is possible by new vapour-liquid-solid epitaxy method. For example, Si nanowires prefer to grow along the <111> direction. Hence if (111) Si wafer is used as substrate, Si nanowires will grow epitaxially and vertically on the substrate. ZnO nanowires prefer to grow along the <00> direction.

The positions of the nanowires can be controlled by the initial positions of the Au clusters or Au thin films. By creating desired patterns of Au using the lithographic technique, it is possible to grow ZnO nanowires of the same designed pattern as they grow vertically only from the region that is coated with Au and form the designed patterns of ZnO nanowire arrays.

2.7 SYNTHESIS OF NANORODS

Recent years enormous progress has been made in synthesizing nanorods with particles size distributions that are within 10% of the mean diameter and frequency within 5%.

Seed-mediated growth method : The reduction of a metal salt in aqueous solution by a strong reducing agent, to make 3–4 nm metallic 'seeds' which are generally spherical is done. The seed reaction is done in the presence of sodiumcitrate to prevent the seeds from aggregating and precipitating. The seeds are then added to a 'growth' solution that consists of fresh metal salt and a surfactant, cetyl trimethyl ammonium bromide CTAB, that directs the growth of nanoparticles into nanorods and nanowires. Growth is initiated by the addition of a weak reducing agent, usually ascorbic acid, that cannot reduce the metal salt on its own at 25°C. Varying the seed to metals salt ratio controls the aspects ratio of the resulting nanorods. Additionally one can use short nanorods as 'seeds' on which to grow longer nanorods. Gold nanorods with diameters of 20nm and aspect ratio short nanorods from 2 to 20, in a controllable fashion. For silver short nanorods with aspect ratio ~ 4 and long nanowires (aspect ratios 50–350). In the case of nanowires, the level control is more limited. The % yield of nanorods compared to spheres is only ~ 20%. However slight changes in reaction conditions, 90% yields are obtained. For gold, final product size and shape are sensitive to experimental conditions such as cleanliness of vessel, temperature and time of growth steps.

Porous alumina membranes with well defined nanochannels are used as hard templates to electrochemically deposit metals. The metal forms nanorods dictated by the channel dimensions and the membrane must be attached to the electrode. Metal nanowires can also be grown via electrodeposition along the step edges of highly-ordered pyrolytic graphite from metal in solution using a STM tip to pulse the voltage in 'activation' 'nucleation and growth' steps.

Many methods of producing nonspherical well defined nanoparticles of various shapes and sizes use a "soft template" (e.g. micro emulsions, polymers or surfactants) that directs nanoparticles growth, in the absence of performed seed. Some of these preparations are performed at higher temperatures in organic solvents with organometallic precursors while other a simple reductions of metals or arrested precipitation reactions in water. Metals, semiconductor and metal oxides are made by these soft solution routes. Semiconductor nanowires are made by heterogeneous seeded methods. This anisotropic growth of a 'nanoparticle in the "soft template" methods involves physical constraints on the part of the template; for example that the size and shape of the resulting nanoparticles made in those matrices. Preferential adsorption of molecules and ions to different crystal facets of the growing nanoparticles leads to different nanoparticles shapes.

Gold nanorods prepared in aqueous solution by the seed-mediated growth method, the growth of the isometric turned "seeds" (diameter of 4nm) in the presence if CTAB results in the initial transformation of Ca. 4% of the seeds into short nanorods, while the remaining crystals increase in size to around 17nm. Once formed the nanorods grow almost unidirectionally in length when immersed in a fresh solution to produce cylindrical penta-twinned particles with high aspect ratios and variable crystal lengths between 100 and 300nm. Because the increase in width is minimal, the elongated crystals have a uniform thickness that is determined by the width of the short nanorods formed in the previous stage of the reaction sequence. At the same time, Ca 6% of the 17nm sized isometric twins are transformed into a new population of short rod shaped nanoparticles while the remaining crystals continue to grow isometrically. Subsequent transfer of the products into fresh reaction solutions reiterates the combination of isometric growth, nanorod elongation and nano rod formation to produce a trimodal distribution in rod widths. The distribution corresponds to 3 types of nanorods with mean widths of 34, 40 and 58nm and decreasing aspect ratios with values between 17–20, 8–11 & 2–3 respectively. Each type can be correlated with the widths of the transferred cylindrical short rod like or isometric nanoparticles respectively which in turn are related to the dimensional and shapes of crystals formed in early stages of the sequential process. Isolation of the higher aspect ratio nanoparticles by centrifugation and precipitation works relatively well. In general, the formation of mixed populations of gold nanoparticles and their associated morphologies and sizes can be rationalized by differences in the delay in the onset of shape anisotropy in the reaction sequence. Clearly, the mechanism responsible for the transformation from isotropic nanoparticles is not highly competitive, although once achieved the crystals rapidly elongate along the common (110) axis suggesting that the process is essentially auto catalytic. As the unidirectional growth rate is high and the onset of the shape transformation process occurs over a extended reaction period, the nanorods originate at different times to produce a marked variation in the particle lengths. In contrast, the width of each particle increases only slowly. This indicates that the (100) side edges are effective blocked from further growth combined with the (111) end faces. The steric and chemical factors of the surfactant play an important role in determining the preferential interactions between the cationic quaternary ammonium head groups and growth sites on the side edges and faces.

Self assembly and designed chemical linkages : While synthesizing and growing two approaches are there. In self assembly the nanorods can be "left to themselves" to order. If one puts lateral pressure on a random 'raft' of rods at the air-water interface. Increasing the concentration of 'hard colloidal' rods in solution would lead to one dimensional or two dimensional liquid crystalline ordering which has been verified. In "designed" assembly, the different crystal faces of the nanorods could be specifically reacted with some reagent to link them in a rational way.

2.8 SYNTHESIS OF NANOCRYSTALLINE MATERIALS

The seminal work carried out in 1970 by Esaki and Tsu is the milestone in the synthesis of nanocrystals. Self assembled nanocrystals has been a field of continued interest.

Synthesis of silver nanocrystals : The synthesis of silver nanoparticles involves the mixing of two reverse miscelle solutions, each with the water content parameter ω(H_2O)/(AOJ) fixed at 40. The total AOT concentration in the first solution is distributed as 30% Ag (AOT) and 70% Na (AOT). The second solution contains Na(AOT) and 0.07M hydrazine. After the solutions are mixed and the Ag particles formed from the reduction of Ag^+ by hydrazine. The addition of dodecanethiol results in the coating of the particles with dodecane thiolate layer and subsequent flocculation. The solution is then filtered to isolate the particles which can be easily redispersed in hexane. After the dodecane thiol extraction step, the size distribution is reduced from 40 to ±30%. Since the poly dispersity is still large size selective precipitation is a well known technique for separating mixtures of copolymers and homo polymers that occurs during the synthesis of sequenced copolymers. This is used for the extraction of nanocrystals. SSP utilizes a mixture of two immiscible solvents each differing in its ability to dissolve the surfactant alkyl chains. For example, alkane thiolatecoated Ag particles are highly soluble in hexane but poorly soluble in pyridine. Thus pyridine is added in steps to a hexane solution containing coated Ag nanoparticles. When the hexane/pyridine volume ratio is ~50%, the solution becomes cloudy indicating agglomerations of the largest Ag particles. The larger particles agglomerate first because of their stronger Vander Waals interactions. This solution is centrifuged and the fraction of solution rich in particle rich in agglomeration is separated. The resulting supernatant is small particle rich. The Ag particle in the agglomerate rich fraction can be redispersed via addition of hexane to form a clear solution. By repeating this, very small particles are got.

Figure 2.7 presents the silver nanocubes. SEM images reveals tiny crystalline particles on the surface. By Tilting by 20^0, the regular cubes are seen. The diagonals

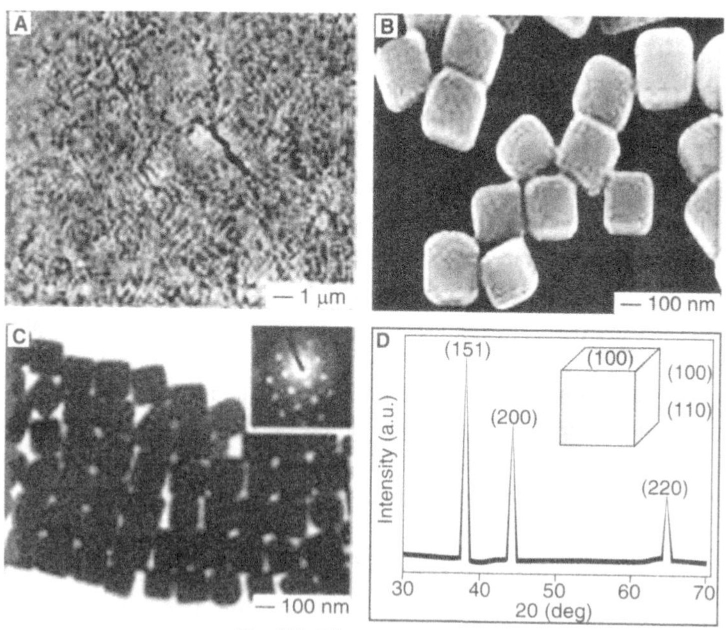

Fig. 2.7 Silver nano cubes.
(A) Low & (B) High magnification SEM images; (B) was taken at a tilting angle of 20; (C) TEM image; (D) XRD pattern.

of the cubes are less than 100 nm. TEM images revealed smaller cubes of silver. XRD patterns revealed (100) (220) and (110) planes.

Synthesis of Ag_2S nanocrystals : Ag_2S nanocrystallites are synthesized by mixing two 0.1M Na (AOT) micellar solutions (hexane solvent) each having the same water content, but with one containing 8×10^{-4}M Na_2S and the other 8×10^{-4}M (AOT). Brownian motion leads to collisions between the two types of micelles and in some cases there is an efficient exchange of water pool contents.

After a few minutes, the reaction of Ag^+ and S^{2-} leads to nano sized Ag_2S particles. Significantly the average size of the particle increases linearly with water content. The standard deviation in the distribution of the particle size is ~30% relative to the average size. Pure alkane thiols are added (1μL/l) to the reverse micelle solutions containing Ag_2S nano crystals. After evaporation of the solvents at 60°C, the resulting solid is washed with ethanol and filtered. The Ag_2S nanocrystals coated with the alkane thiols are then dispersed which does not allow for the extraction of the largest particles (10nm) from the reverse micelles. The extraction engenders a size selection resulting in a smaller average diameter as well as a decrease in the relative distribution of sizes from 30% to 14%.

Semiconductors : Synthesis of these compounds is of technological interest. There are mainly two major routes viz., physical vapour deposition and chemical methods.

Gas-phase condensation technique is used. In this method, metal atoms are evaporated in an evaporation chamber by conventional evaporation technique from an appropriate boat. The evaporated particles are then transferred to another chamber (deposition chamber) through a transfer pipe where the evaporated particles in aerosol state suffer multiple collisions and get fragmented before getting deposited on a cooled substrate surface.

In laser ablation technique, a high power pulsed laser is utilized to evaporate materials from the surface of the source material kept in an inert gas chamber. Energy of the incident laser beam has to be selected for individual material. Clusters of semiconductors are made.

Sputtering is used where high melting point materials are to be prepared. Sputter deposition uses a magnetron unit. The vacuum chamber is made of cylindrical stainless steel. There are appropriate stainless steel vacuum coupling through which requisite vacuum gauges, gas inlets, thermocouples etc. could be attached to the system. The sputtering target could be prepared by pressing the target material (99.999% pure) into aluminum disc of suitable diameter and thickness. The disc is mounted in a specially designed aluminum holder. An appropriate magnet assembly can be housed into the target holder to transform the system to a magnetron sputtering system. The target assembly expect the target surface is covered outside by a cylindrical teflon rod; The target holder is attached to the top of a stainless steel plate after insulating it by a Teflon sheet. During sputtering the plasma is confined with in a cylindrical zone. The substrate holder is a heavy copper plate (~10mm thick) placed at a distance of 2.5cm below the target. The substrate holder also serves as a cold finger (cooled by liquid N_2). Intermediate temperatures can be attained by heating the copper plate with appropriate heating elements passed through the holes laterally drilled through the copper plate. Films are deposited by sputtering the target in Ar plasma. Before starting the actual deposition the target is pre-sputtered for appropriate duration by using shutter covering the substrate. The shutter is also used to control the time of deposition. The Ar gas pressure is varied within 10–40pa. The flow rate is controlled by a mass flow meter. Due to the high gas pressure of the system, the particles ejected from the target undergo many collisions with the Ar gas atoms and subsequently they nucleate and grow by rapid condensation on the substrate to form ultrafine particles.

Dip coating and precipitation are two common techniques. Dip coating method has a chemical bath. A condenser with a water seal can be added at the top to prevent the loss of volatile materials. The sample hold can be lifted outside the bath fluid whenever required. Direct heating of the bath fluid is also avoided and the bath is placed in an oil bath which can be heated by a magnetic, stirrer cum heater. The semiconductor for example CdS films are deposited on to sodalime glass with thiourea ammonia aqueous solution and different cadmium salts like $CdCl_2$ and CdI_2. Films are deposited at various temperatures (60–80°C), thiourea content and pH values of the bath. pH values and the temperature of the bath solution are constantly monitored during the whole process. The thickness of the film is determined usually by Stylus method and is 15–40nm. The chemical reactions leading to CdS deposition are :

$$CdA_2 + 2NH_4OH \leftrightarrow Cd(OH)_2 + 2NH_4A$$

$$Cd(OH)_2 + 4NH_4OH \leftrightarrow [Cd(NH_3)_4]\,(OH)_2 + 4H_2O$$

$$[Cd(NH_3)_4]^{2+} + S^{2-} \leftrightarrow CdS + 4NH_3$$

where, $A = Cl^-$ or I^-.

Figure 2.8 shows the semiconductor nanoparticle CdS synthesis. Cadmium and sulphide ions are mixed to form CdS. The nanoparticles are precipitated under

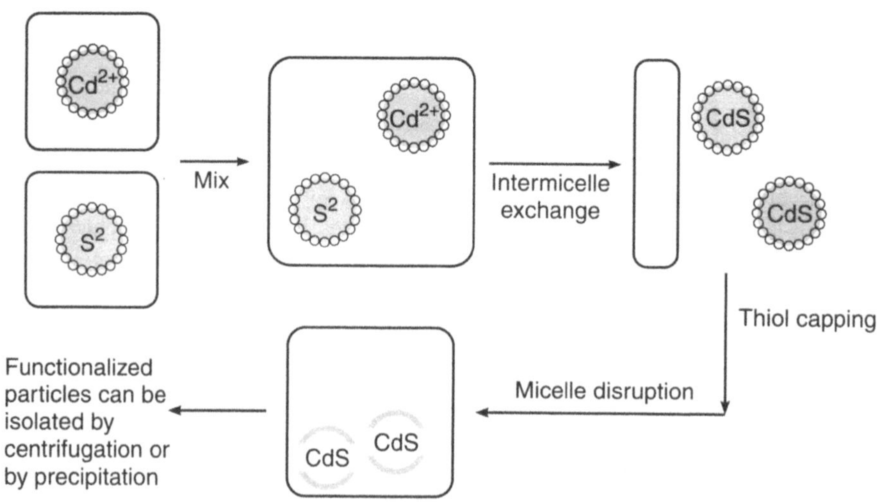

Fig. 2.8 Semiconductor nanoparticle synthesis.

starving conditions, a large number of nucleation centres are formed by vigorous mixing of the reactant solutions. If concentration growth is kept small, nuclei growth is stopped due to lack of material. Against aggregation, nanoparticles are stabilized and one of the common stabilizer is thiols (thioethanol, thioglycerol, mercapto ethyl amines etc). The micelles are disturbed, particles are precipitated or centrifuged.

Precipitation method for making CdS nanoparticles from an aqueous solution containing $CdCl_2$, thiourea and NH_3 is given below. Measured quantities of $CdCl_2$ and NH_4Cl are added to an aqueous bath at 65°C and dissolved by continuous magnetic stirring. The capping agent, thiophenol is then added to the solution. The pH of the mixture is raised to 7.5 by adding NH_4OH. Precipitation is achieved by adding a measured quantity of thiourea and the reaction is allowed to proceed for ~30 minutes. The presence of thiophenol arrests the agglomeration of the CdS particles during precipitation. Since thiophenol forms a coating on the surface of the CdS nanoparticles, it effectively reduces the heterogeneous nucleation of new particles on the surface of the existing ones thus reducing the precipitation rate. The precipitate is separated from the aqueous solution by centrifuging at 8000rpm. The separated nanoparticles are repeatly washed in methanol to remove traces of thiophenol and unreacted chemicals. After drying some portion of the CdS sample is dispersed in DMSO to form a stable suspension for spectroscopic studies. The dispersion is ultrasonically agitated to break up any CdS agglomerates, which might have formed during centrifuging and drying. The chemical reaction that results in the precipitation of CdS nanoparticles are

$$Cd(NH_3)_4^{2+} + SC(NH_2)_2 + 2OH^- \rightarrow CdS + H_2CN_2 + 4NH_3 + 2H_2O$$

2.9 SYNTHESIS OF OXIDE NANOPARTICLES

There are three popular methods to synthesis: (a) Mechanical, (b) Physical, and (c) Chemical methods.

(a) Mechanical method : Ball milling is the simplest and easiest method to achieve particles at submicron level. High purity oxide materials in proper portion are mixed together. Mixing is often performed by wet milling in a rubber lined pot using stainless steel or alumina balls; the mixture is dried and calined. The hard cake obtained during calcinations process is ball milled once again for a long time to get fine particles. Though the process is simple it has some serious limitations such as lack of control on chemical purity, homogeneity and particles size distribution. The process is lengthy and takes about a month to bring the particles size to submicron level.

(b) Physical methods : Solid phase transition, spray drying and liquid drying methods are used.

Irreversible phase transformation is accompanied by marked reduction in particle size. The reduction in particle size is proportional to change in unit cell volume at transition. In order to control the particle size, the calcinations temperature needs to be restricted just above the phase transformation. It helps to avoid the growth of the freshly nucleated phase. The method is applicable if suitable meta stable polymorph is available. In certain cases it could be an alternative to wet chemical route.

Spray drying is used to synthesis oxide nanoparticles in this method. The solution to be dried is fed to the eye of the centrifugal atomizer disc. Due to rotation of the disc at high speed, the solution is thrown out in the form of a fine mist. Hot dried air is simultaneously introduced through an annular opening into the drying chamber. This hot air mixed continuously with the mist of atomized liquid and an instantaneous evaporation of the liquid takes place. The non-volatile part is left in the form of dry powder. The powder falls down towards the outlet aperture in the conical bottom of the chamber and from there it is carried into the dynamic cyclone separator by the air current. Then the powder falls down into a container and the drying air is discharged from the top of the cyclone. The dry powder is then calcined to get the required phase. In this technique particle size is controlled by speed of the disc, solution composition and calcinations temperature. Special care is required to preserve the particle size if the material is hygroscopic. In case of multi cation systems to obtain reacted phase the reactants have to undergo solid state diffusion as there is no complex formations.

Liquid drying method in which the stiochiometric aqueous solution is sprayed as a broken stream of droplets into the vertex of swirling bath of hygroscopic liquid. Acetone is used as a hygroscopic liquid. Acetone-water mixture is removed from the dry powder by vacuum filtration. Since no formation of complex occurs in multi cation systems, solid state diffusion process is unavoidable to achieve the reacted phase. Particles size is decided by the solution concentration and calcinations temperature.

(c) Chemical methods : Co-precipitation route in which the constituents are weighed in stoichiometric proportion and dissolved to get the stock solution. All the cations are then co precipitated to achieve the complex formation by selecting suitable precipitating agent and pH. In case of oxides, the precipitate could be in the chemical form of hydroxide, carbonate, oxalate etc. so that after decomposition is converted into oxide. In this process the particle size is determined by the solution concentration, rate of precipitation, calcinations temperature. However, during the drying of the precipitate, there is a possibility of agglomeration of the particles and hard cake obtained in this step needs ball milling.

Sol-gel method is one of the well proven methods to produce fine particles with narrow size distribution and controlled chemical composition at comparatively low temperatures. In this process dispersion of the particles of the metal compound (sol) usually in an aqueous phase is first prepared and then converted to gel particles. Gelation is achieved by means of chemical dehydrating agent with surfactant. 2-ethyl hexanol and a surfactant is commonly used. In principle particles size achieved at the precipitation state is maintained during sol and gel formation. Since agglomeration is avoided, the process gives better control on particle size and size distribution. Column technology is used if gelated material is required in spherical shape. The gel is then dried and calcined to get reacted product.

Sol-gel auto combustion method is a novel way and a unique combination of the combustion process and the chemical gelation process. The process exploits the advantage of cheap precursors and simple preparation resulting ultra fine, homogeneous highly reactive powder. Usually metal nitrate citrate and PVA (to control size) and dissolved stoichiometrically. The water in all of the citrate solutions are evaporated at 0–90°C. After evaporation of 2/3 of the water, a clear solution would be turbid. Sometime the dehydrated solutions are transformed into gels. The As prepared sol-gel is dried homogeneously at 180°C into dry, loosened and foamy structural gels. Then the dry gels are ignited at 600°C in a preheated furnace. After 10–15 minutes powders are formed.

Solvothermal method is a recent one for making MgO, ZnO, Al_2O_3, Fe_2O_3 and ZrO_2. When materials are irradiated with microwaves they absorb energy from the microwave field. The power absorbed P is $P = 2\pi f\ \varepsilon_o\ \varepsilon'\ \tan\delta\ |\varepsilon_i|^2$. The frequency, f and the amplitude $|\varepsilon_i|$ of the microwaves are instrumental parameters while the dielectric constant ε' and the loss tangent, tan δ are materials parameters. Therefore in a given instrumental setup $|\Delta\varepsilon_i|$ and f held constants. P is determined by ε' and tan $\delta.\varepsilon'$ is generally high for materials containing molecules or complex ions of high dipole moment and therefore high molar polarization. When a mixture of two material A and B of different dielectric constant ε_A' and ε_B' is exposed to microwave field the material with higher values of ε' absorbs energy preferentially and gets heated rapidly compared to the other. For the microwave decomposition of a material in a mixture the differential absorption is used. A precursor material is chosen such that it has a high value of ε' and decomposes by preferential microwave absorption to yield the desired oxide. Thus if a metal organic or an inorganic complex salt, of reasonably high value of dielectric constant that yields of desired oxide upon decomposition, is dissolved in a suitable liquid medium of low dielectric constant and irradiated with microwaves, one can expect the formation of an oxide. The oxide generally forms as

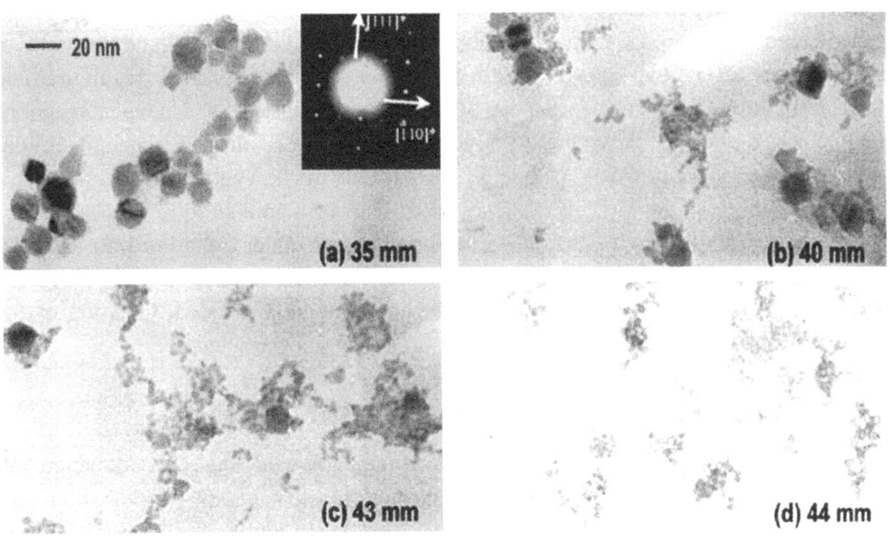

Fig. 2.9 Fe_2O_3 Nanocrystals—solvothermal method.

a suspension in a chosen liquid. It is better to choose a precursor which has high values of both ε' and tan δ, but tan δ values are not often readily available and are highly temperature dependent. This may result in a disadvantage occasionally if tan δ decreases with temperature (as in a post transition region) in the region where ε' increases. The liquid should not decompose before its normal boiling temperature. The oxide product should be prevented from going into large particles. This is accomplished by the addition of a suitable capping agent to the initial solution. It is necessary to control microwave power. Solvothermal method was used to get Fe_2O_3 nanocrystals sizes of 20–30 nm and shown in **Figure 2.9**.

Emulsion Method

When two immiscible liquids are mixed together, emulsion is formed. However to get micro emulsion (stable and transparent system) the globule size of the dispersed phase needs to be $< 1000A°$. A micro emulsion is basically a three phase system, example oil, water and a surfactant. There are four main stages: (1) Formation of micro emulsion; (2) Precipitation of inorganic phase; (3) Separation of the precipitate; and (4) Drying and calcinations.

An emulsion combustion method (ECM) uses a flame as a heat source. The temperature of the flame is less than that in flame spray pyrolysis. (FSP) when dissolving a precursor with organic solvent (e.g. ethanol and octane) instead of water. The spray is ignited to form a self sustaining flame, which behaves as a reactor (FSP). In ECM the reaction occurs in the liquid phase as in spray pyrolysis. ECM is used to make nanoparticles of oxides.

A micro-emulsion is generally defined as a system composed of a mixture of an aqueous phase, an oil phase, surfactants and co-surfactants. After these components are mixed, a transparent solution that is thermodynamically stable and optically isotropic is obtained. The concept of the micro-emulsion process involves thoroughly dispersing the water phase into the oil phase using a surfactant, nano sized micro-emulsion droplets (reverse micelles) are formed. Each tiny droplet acts as an independent micro-reactor for producing the desired compounds and provides a confinement effect that prevents particle to particle contact. As a result particle nucleation, growth and coagulation can be significantly constrained. Chemical reactions take place rapidly in these micro-cavities containing the desired reactant colloid allowing nanosized powders with high crystallinity to be synthesised.

Sol-emulsion-gel process is also used. The principle of this process is the dispersion of a sol-containing the desired constituent under agitation into a water-immiscible organic liquid for low dielectric constant. To stabilize the dispersed sol droplets addition of an amphiphilic surface active agent is needed. Gelation of the sol droplets with a suitable gelling agent produces gel particles.

2.10 SYNTHESIS OF NANOINTERMETALLICS

Over the past one decade, a considerable volume of work has been done on the synthesis of nanointermetallic compounds through mechanical alloying. High energy ball milling is used. In this method the energy transfer during ball milling takes place by shear or impact of the high velocity balls with the powder. An attritor is the first high energy ball mill used for mechanical alloying (MA). Where the movement of the balls and powder is achieved by the horizontally rotating impellers attached to a vertical shaft and set progressively at which are traditionally used in mineral processing can also be used for MA, if their diameters are sufficiently large (of the order of meters) and of mills are operated close to the critical speed beyond the balls stick to the inner walls of the mill. For large scale production, tumbler mills are more economical when compared to the attritor and other high energy ball mills. The uni ball mill involves a controlled movement of grinding balls that is confined to the vertical plane by the cell walls and adjustable external magnetic field. The mill can run in high energy (impact) or in low energy (shear) mode. In addition to the above mills, rod mill, vibrating mills are also known. A modified electric discharge assisted mechanical milling technique is developed where materials react in a gas atmosphere under an electric discharge. Application of low current high voltage electric impulses results in a faster reactions and new synthesis and processing routes. Glow (cold) discharge milling is found to promote solid-gas reaction where as spark (hot) discharge milling promotes fast fracturing, recrystallization, mineral reduction and solid-solid reactions.

The kinetics of phase formation appears to depend on the energy transferred to the powder ingredients from the balls during milling. The energy transfer is governed by various parameters e.g. the design of the mill, milling speed, type, size and size distribution of the balls, ball to powder weight ratio (BPR), extent of filling of the vial, temperature of milling, grinding media, milling atmosphere and the duration of milling.

Majority of the MA work makes use of stainless steel or hardened chrome steel grinding materials. However, these can introduce iron contamination into the milled powder due to continuous wear caused by the impact and friction involved in the process. Contamination can cause profound influence on the MA process as well as the product quality. The kinetics of nanocrystal formation during milling e.g. in TiNi is a function of the milling temperature.

Nanointermetallics of aluminides and silicides are popular. They are synthesised by MA. $NiAl_3$, NiAl and Ni_3Al through MA are known. Though a considerable extension of phase fields in the Ni_3 Al, Ni Al and even in the line compound $NiAl_3$ are observed under intensive milling, Ni_2 Al_3 and Ni_3 Al_3 phases are found metastable under similar conditions. In a marked contrast to ~348K during MA of Ni aluminides under air, the formation of NiAl in spex 8000 mill under Ar atmosphere is accompanied by an exothermic reaction within a short duration of milling following an interruption after 2h of continuous milling. Ni_{50} Al_{50} is prepared by low energy milling. Ti-Al intermetallics can be achieved on adopting a 2-stage 'mechanically activated annealing process' involving intense ball milling and subsequent annealing. MA of Fe_{20} Al_{80} and Fe_{25} Al_{75} blends in planetary ball mill has shown $FeAl_3$ formation. The formation of $FeAl_3$ at non stoichiometric composition may be an indication of an extended. $FeAl_3$ phase field though any possible Al loss due to oxidation may be responsible for such observation. MA of Ferich compositon e.g. Fe 25 at% Al which lies in the Fe_3Al phase field is found to produce FeAl. The formation of FeAl during MA in preference to $AlFe_3$ can be attributed to its more negative ΔH_f (–31.8KJ/mol) when compared to that of AlFe3 (–18KJ/mol).

Silicides of Ni, Fe, Ti and Mo find applications in micro electronics and electrical engineering. MA of Ni_{100-x} Si_x (x = 25, 28, 33, 40 & 50) at a speed of 642rpm yields crystalline phase formation of Ni_3 Si, Ni_3 Si_2 and Ni Si compositions and an amorphous phase in Ni_5 Si_2, Ni_2 Si compositions. MA of elemental Fe_{33} Si_{67} blend gives the formation of a mixture of low temperature tetragonal α Fe Si_2, high temperature orthothombic β Fe Si_2 and the cubic Fe Si phases. MA on Fe_{100} Si_x (x=5, 25, 37.5 & 50) yields the formation of FeSi, Fe_2Si, Fe_5Si_3 and Fe Si_2 respectively. Fe-6.5wt % Si is a solid solution. β Fe Si_2 phase is found for Si >70 at % and $\alpha+\beta$ $FeSi_2$ at 50 at% < Si < 70at%. MA of Ti-Si system yields amorphous compounds. Mo-silicides is synthesized by MA of Mo_{33} Si_{67} at 280 rpm disc speed. The low temperature as well as the high temperature phases viz., α and β $MoSi_2$ evolved during MA in spex mill; high energy ball milling results in a self propagating high temperature. (SHS) leading to stable α $MoSi_2$, while low energy milling leads to metastable β $MoSi_2$ with out any such reaction.

2.11 SYNTHESIS OF NANOCOMPOSITES

In situ synthesis of nanocomposites by MA is advantageous as it cuts down the production steps and assists better inter mixing of the constituents and leads to enhanced properties. Composites of Pb-Al, Fe-Cu, Ni-Ag, Cu-Nb are known. No grain growth occurs in the matrix close to its melting point when Cu-Mg are reinforced with nanocrystalline Al_2O_3 by MA. This nanocrystals are formed in an amorphous Ti-Al matrix. A solid reaction leading to the formation of AlN and AlB_2 occurred during further processing of these nanocomposites. TiAl-Ti_5 Si_3 nanocomposites are made by thermally activated MA. There is a formation of these nanocomposites to the crystallization of amorphous phase obtained by MA in Ti-Al-Si system. These compounds are stable. Nb Al_3 – NbC nanocomposites are obtained by MA of Al and Nb powders. The formation of Cu-$NbAl_3$ nanocomposite by the co-deposition of nanocrystalline Nb Al_3 and copper is known. Nanocrystals of Ti_5 Si_3 dispersion synthesized by Ti Al and Si give enhanced microhardness. Nano WC-Co and ultra fine Ti (CN) based cermets containing nano-Ni binder is found to have enhanced tribomechanical properties due to inter diffusion.

2.12 SYNTHESIS OF POLYMERIC NANOFIBERS

Drawing : Nanofibers have been fabricated with citrate molecules through this process. A laboratory method involves a freshly prepared aqueous solution of 50mL of chloroauric acid (0.01% w/v). This is boiled under reflex and 1.75mL of an aqueous solution of sodium citrate (1% w/v) is added. A ml droplet of the chloroauric acid/sodium citrate solution is deposited on a SiO_2 surface and allowed to evaporate.

The solution becomes more concentrated at the edge of the droplet due to capillary flow after a few minutes of evaporation. A micropipette with a diameter of a few micrometers is dipped into the droplet near the contact using a micromanipulator. The micropipette is then withdrawn from the liquid and moved at a speed of ~100 μm/sec resulting in a nanofiber being pulled. The pulled fiber is deposited on the surface by touching it with the end of the micropipette. Fibers are also drawn from a pipette containing the chloro-auric acid/sodium citrate solution with the help of the micropipette and the micromanipulator. The drawing of nanofibers are repeated several times on every droplet. The viscosity of the material at the edge of the droplet increased with evaporation. At the beginning of evaporation to the drawn fiber breaks due to Raleigh instability. During the second stage of evaporation nanofibers are drawn. In the final stage of evaporation of the droplet the solution is concentrated at the edge of the droplet and breaks in a cohesive manner. Thus drawing a fiber requires a viscoelastic material that can undergo strong deformations while being cohesive enough to support the stresses developed during pulling. This process is considered as dry spinning at a molecular level.

Template synthesis : A simple template synthesis approach is used to extrude aligned Poly acrilonitrile (PAN) nanofibers. It consisted of the following steps: PAN is dissolved in DMF (dimethyl formamide) at 70°C and stirred to form a 18 wt% precursor solution. A solidifying solution consists of a mixture of 40 wt% DMF and 60 wt% deaerated water is prepared. A clean and dry aluminium oxide membrane is made anode. It has an average pore diameter of 103.4nm and a pore density of 6.43×10^9 Pores/cm^2. This template is placed on a compact polytetra fluoro ethylene film and the precursor solution is poured on the template. The precursor is extruded as PAN nanofibers into the solidifying solution under a water pump pressure of 0.1Mpa. The solidified PAN nanofibers are removed, washed with deionized water several times and dried in air. It is found that the surface of the PAN nanofibers is super hydrophobic without any modification by materials of low surface energy. Templates can also be designed with different pore diameters.

Phase separation : An extracellular matrix is created using nanofibers from poly (L-lactic acid) (PLLA) using the process of phase separation. It consists of the following steps :

(1) PAN is dissolved in DMF at 70°C and stirred to form a 18 wt% precursor solution.

(2) A solidifying solution consisting of a mixture Of 40 wt% DMP and 60 wt% deaerated water is prepared.

(3) The precursor is extruded as PAN nanofibers into the solidifying solution under a water pump pressure of 0.1M Pa.

(4) The solidified PAN nanofibers are removed, washed with deionised water several times and dried in air.

Another method also involving phase separation is given below : An extracellular matrix has been created using nanofibers from poly (L-Lactic acid) (PLLA). Phase separation consists of the following 5 steps.

Polymer dissolution : Tetrahydrofuran (THF) was added to PLLA in a flask to make a solution with the required concentration (1% w/v to 15% w/v). The solution is stirred with a magnetic stirrer at 60°C for 2 hours to produce a homogeneous solution.

Gelation : 2ml of the solution at 50°C is poured into a Teflon vial and transferred to a refrigerator set to a gelation temperature (18°C to 45°C) which is chosen based on the PLLA concentration. Once the gel is formed, it is kept the gelation temperature for 2 hours.

Solvent extraction : The vial containing the gel is immersed in distilled water for solvent exchange and the water is changed 3 times a day for 2 days.

Freezing : The gel is removed from water, blotted with filter paper and then transferred to a freezer at –18°C and kept for 2 hours.

Freeze drying : The frozen gel is transferred into a freeze-drying vessel and freeze dried at –55°C under a vacuum of 0.5mm of Hg for 7 days.

Nanofibers matrices from PLLA and PLLA-poly caprolactone (PCL) blends of blend ratios of 80:20 and 50:50 have been also fabricated. Proper control of cooling to a temperature lower than the T_g of the polymer in solution result to a get a gel.

Self assembly : Nanofibers are constructed from a polystyrene-block-poly (2-cinnamoyl ethyl metha acrylate) di block copolymer or PS-b-P CEMA through the self assembly process. The self assembly process consists of the following 3 steps.

Preparation of the cylindrical phase : A macromolecule consisting of 2 linear polymer chains joined together in a head to tail manner is called a d-block copolymer $(A)_n$–$(B)_m$. The blocks segregate from one another in bulk into A and B domains due to their incompatibility. The shape of A or B domain varies with the relative n and m values and the temperature. As the volume fraction of B increases the shape of the B domain changes from spheres (~17%) to cylinders (~28%) gyroid's (~28%) and finally lamella (~50%). The bulk PS-b-PCEMA film is annealed at 110°C for 3 weeks so that the PCEMA phase existed as cylinders dispersed in the PS matrix.

Cross linking : The annealed film is irradiated with uv light for 40 minutes on each side to achieve uniform cross linking of the PCEMA cylinders across the sample.

Dissolution : The irradiated film is immersed in THF for 24 hours and the continuous PS phase dissolved to disentangle the nanofibers. The supernatant is concentrated and precipitated into CH_3OH to yield nanofibers with a PCEMA core and a PS shell.

Polymer-Fe_2O_3 hybrid nanofibers from the triblock copolymer polystyrene-block-poly (2.Cinnamoyl ethyl metha crylate)-block-poly (tert-butyl acrylate) or PS-b-PCEMA is consonated. The construction involves :

Preparation of the cylindrical phase : The bulk PS-b-PCEMA-b it BA film is annealed at 110°C for 3 weeks so that the PCEMA and Pt BA phases existed as cylinders with Pt BA cores and PCEMA shells dispersed in the PS matrix.

Cross linking : The annealed film is immersed in THF for 24 hours and the continuous uncross-linked PS phase is dissolved to disentangle the nanofibers. The supernatant is concentrated and precipitated into CH_3OH to yield nanofibers with a PtBA core, PCEMA middle layer and a Ps corona.

Cleaving : Nanofibers with poly (acrylic acid) or PAA cores are obtained by cleaving he tart butyl groups from Pt BA through the treatment of the fibers with $(CH_3)_3SiI$ and then decomposing the excess with CH_3OH containing 2% water.

Loading : The PAA cores are loaded with γ-Fe_2O_3 resulting in supra magnetic block copolymer\inorganic hybrid nanofibers. Nanofibers are produced from poly (4-Vinyl pyridine) – block PS or P4 VP-b-PS diblock copolymer using the self assembly process without the use of cross linking. The method consisted of :

Hydrogen bonding : P4VP (PDP)-b-PS diblock copolymer comb-coil supra molecules were produced by H bonding P4VP-b-PS to a stoichiometric amount of 3-n pentadecylphenol (PDP). Selection of appropriate lengths of the P4VP(PDP) and PS blocks resulted in a microphase separated cylindrical morphology of Ps cylinders in P4VP (PDP) matrix.

Dissolution : The supra molecules are dissolved in to yield nanofibers with a PS core and a corona of P4VP.

Electro spinning : It is a process that creates nanofibers through an electrically charged jet of polymer solution or polymer melt. The process in its simplest form consisted of a pipette to hold the polymer solution, two electrodes and a direct current supply in the KV range. The polymer drop from the tip of the pipette is drawn into a fiber due to the high voltage. The jet is electrically charged and the charge causes the fibers to bend in such a way that every time the polymer fiber looped, its diameter is reduced. The fiber is collected as a web of fibers on the surface of a grounded target. Electro spinning method to get polymeric nanofibers illustrated in **Figure 2.10**.

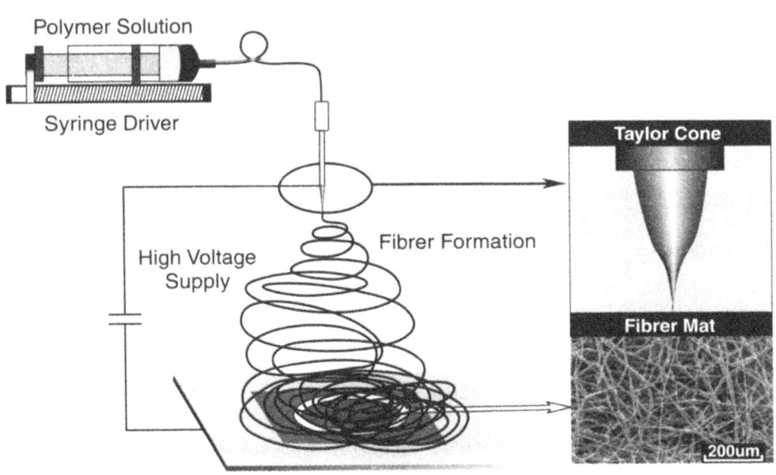

Fig. 2.10 Electro spinning method.

Electro spinning method has some unique features like :

1. Availability of suitable solvent for dissolving polymer.
2. The vapour pressure of the solvent should be suitable so that if it evaporates quickly enough for the fiber to maintain its integrity when it reaches the target but not too quickly to allow the fiber to harden before it reaches the nm range.
3. The viscosity and surface tension of the solvent must neither bee too large to prevent the jet from forming nor too small to allow the polymer solution to drain freely from the pipette.
4. The power supply should be adequate to overcome the viscosity and surface tension of the polymer solution to form and sustain the jet from the pipette.
5. The gas between the pipette and grounded surface should not be too small to create sparks between the electrodes but should be large enough for the solvent to evaporate in turn for the fibers to form. **Table 8** provides a list of polymers that can be converted to nanofiber and the processing methods.

TABLE 8. List of the polymers converted into nanofibers and processing methods.

Process	Material	Solvent
Drawing	Sodium citrate	Chloroauric acid
Template synthesis	Polyacrylonitrile	DMF
Phase separation	PLLA	THF
	PLLA-PCL blends	THF
	PCEMA core, PS shell	THF
	PAA-ϒ-Fe_2O_3 core-PCEMA	THF
	middle layer, PSCorona, P_4VP	$CHCl_3$
Self assembly	Corona, Peptide-ambhiphile	
	Polyimides	Phenol
	Polyamic acid	m-cresol
	Polyetherimide	methylenechloride
	Polyaramid	H_2SO_4
	Polygamma benzyl glutamate	DMF
	Poly(p-Phenylene tetra phthalamide)	H_2SO_4
Electro-spinning	Nylon-6-polyimide	HCOOH
	Polyacrylonitrile	DMF

Polyethylene-terephthalate	Trifluoroacetic acid
Nylon	
Polyaniline	Dichloromethane
DNA	H_2SO_4
Polyhydroxybutyrate	Water
Valerate	$CHCl_3$
PLLA	$CHCl_3$ mixed with
Poly (D,L-Lactic acid)	Methylene
PEO	chloride/DMF
PMMA	Water
PU	Toluene
	DMF

CHAPTER 3

Characterization of Nanomaterials

Since 1957, after the proclaim of Moore's laws, the continual advancement in technology resulted the design factor changes in chip making. One of them has been the progressive ability to fabricate smaller and smaller structures and another has been the continuous improvements in the precision with which such structures are made. Lithography makes use of a radiation sensitive layer to form well defined patterns on the surface. Molecular beam epitaxy is used to grow one crystalline material on the surface of another. Nanotechnology revolution involves improvement of old and introduction of new instrumentation systems for evaluating and characterizing nanostructures.

3.1 STRUCTURE OF NANOMATERIALS

In order to understand the size, one can compare water molecules, protein and DNA molecules: A water molecule has 3 atoms, protein molecules have 1000 of atoms and DNA molecules have millions of atoms. Most of the nanostructures are crystalline where in thousands of atoms have a regular arrangement in space called a crystal lattice. This lattice can be described by assigning the position of atoms in a unit cell so the overall lattice arises from the continual replication of this unit cell throughout space. There are 17 possible types of crystal structures called space groups meaning 17 possible arrangement of atoms in unit cells in 2 dimensions and these are divided between the 4 crystal systems. The most efficient way to arrange identical atoms on a surface is hexagonal system.

In three dimensions, the situation is much more complicated and there are now 3 lattice constants a, b and c for the 3 dimensions x, y, z with respective angles α, β and γ between them. (α is between b and c and so on). There are 7 crystal systems in 3 dimensions with a total of 230 space groups divided among the systems. The objective of a crystal structure analysis is to distinguish the symmetry and space group to determine the values of the lattice constants and angles and to identify the positions of atoms in the unit cell.

Certain special cases of crystal structure are important for nanocrystals such as those involving simple cubic (SC), body centered cubic (BCC) and face centered cubic (FCC) unit cells. Another important structural arrangement is formed by stacking planar hexagonal layers which for a monoatomic (single atom) crystal provides the highest density or closest packed arrangement of identical spheres.

TABLE 1. Crystal systems and space groups.

Dimensions	System	Space Groups
2	Oblique	2
2	Rectangular	7
2	Square	3
2	Hexagonal	5
3	Triclinic	2
3	Monoclinic	13
3	Orthoclinic	59
3	Tetragonal	68
3	Hexagonal	27
3	Trigonal	25
3	Cubic	36

If the third layer is placed directly above the first layer, the fourth directly above the second and so on in an A-B-A-B—type sequence, the hexagonal close packed (HCP) structure results. If on the other hand this stacking is carried out by placing the third layer in a third position and the 4th layer above the first and so forth, the results is an A-B-C-A-B-C—a sequence and the structure is FCC. The latter arrangement is more commonly found in nanocrystals.

Epitaxial films prepared from FCC or HCP crystals generally grow with the planar close packed atomic arrangement. FCC cubic crystals tend to expose surfaces with this same hexagonal two dimensional atomic array. Silicon has a diamond structure with a = 5.43 Å. A silicon nanocube 10nm on a side has 6250 unit cells. Each nanocube face has 340 unit cells per face and the total surface area has 10% of total number of atoms compared to 0.03% in bulk silicon film of 1μM thickness on a 10 cm^2 area. Some properties of nanostructures depend on their crystal structures while other properties such as catalytic reactivity and adsorption energies depend on the type of exposed surface.

3.2 X-RAY DIFFRACTION

To determine the structure of a crystal, and there by ascertain the positions of its atoms in the lattice, a collimated beam of X-rays, electrons or neutrons is directed at the crystal and the angles at which the beam is diffracted are measured. Now X-ray diffraction is presented in detail. The wavelength λ of the X-rays expressed in nm is related to the X-ray energy E expressed in the units KeV as :

$$\lambda = \frac{1.240}{E}\ \text{nm}$$

The regular arrangements of atoms in the cleavage planes of a crystal might provide a grating element suitable to diffract X-rays. Thus the crystal might serve as a three dimensional space grating where as the ordinary optical grating is a two dimensional plane grating. W.L. Bragg discovered that crystals when the X-rays are incident on their surface nearly at glancing angles. Glancing angle or Bragg angle is the angle between incident ray and crystal plane.

Thus Bragg's law aims at devising the conditions for X-rays reflected by the crystal plane to cause constructive interference. When a beam of monochromatic X-rays incident at a glancing angle 'θ' on a set of parallel planes of a crystal, the beam is partially reflected at the successive layers rich in atoms. Ray no.1 is reflected from atom A is plane 1 where as Ray No.2 is reflected from atom B lying in plane 2 rays will be in phase or out of phase with each other. This will depend on their path difference. This path difference can be found by drawing perpendiculars AM and AN on ray No.2. Since the two rays travel the same distance from A and N onwards it is obvious that ray no.2 travels an extra distance = MB + BN; Hence the path difference between the two reflected beams = MB + BN = $d \sin \theta + d \sin \theta = 2d \sin \theta$, where d is the interphase spacing i.e. vertical distance between two adjacent planes belonging to the same set. The two reflected beams will be in phase with each other if this path difference equals an integral multiple of λ and will be out of phase of it equals on odd multiple of $\lambda/2$. Hence the condition for producing maxima (or) constructive interference becomes $2d \sin \theta = n\lambda$ where n = 1,2,3 for first order, second order and third order maxima respectively. This is Bragg's law.

Now consider the first order spectrum from three planes one after the other reflecting a given monochromatic X-ray beam of wavelength 'θ' at the glancing angles $\theta_1, \theta_2, \theta_3$. From Bragg's law, $2\ \theta \text{ Sin } \theta = \lambda\ (n = 1)$ $1/d = 2 \text{ Sin } \theta/\lambda$; Hence $\lambda = 2d_{100} \text{ Sin } \theta_1 = 2d_{110} \text{ Sin } \theta_2 = 2d_{111} \text{ Sin } \theta_3$; $1/d_{100} : 1/d_{110} : 1/d_{111} = \text{Sin } \theta_1 : \text{Sin } \theta_2 : \text{Sin } \theta_3$. In the case of cubic lattice, there are 3 lattice types. The ratio $1/d_{100} : 1/d_{110} : 1/d_{111}$ are different for each lattice type. These are $\sqrt{1} : \sqrt{2} : \sqrt{3}$ for a simple cubic, $1 : 1/\sqrt{2} : \sqrt{3}$ for BCC and $1 : \sqrt{2} : 3/\sqrt{2}$ for FCC lattice type. Hence knowing the values of glancing angles ratio, the ratio of inter planar spacing can be determined even without knowing the wavelength of X-rays. From that ratio, the type of lattice can be identified.

By knowing the structure of the cubic lattice, the number of atoms / molecules in the unit cell may be calculated. Immediately one can calculate the actual value of lattice distance (or) lattice constant from a knowledge of the molecular weight, density of the crystal and the Avagadro number. Let 'a' be the side of the unit cell. It is also equal to the distance between two like atoms or ions. Hence 'a' is called lattice distance or lattice constant. If 'ρ' is the density of the crystal, then the mass of the unit cell is ρa^3 where a^3 is the volume of the unit cell is. Now $\rho a^3 = nm/N$ where (nm/N) = mass of the unit cell ; n = number of atoms /molecules in the unit cell; M = atomic weight or molecular weight of the crystal ; N = Avagadro number.

From the values of λ and θ, one can calculate the value of inter planar distance "d". 2d $\sin\theta = \lambda$ for first order spectrum, therefore, $d = [\lambda/2\sin\theta]$, in the case of cubic structure, $d = a/(h^2 + K^2 + l^2)^{1/2}$ where a = lattice distance or lattice constant. It is the distance between two like atoms/ions (h,k,l) are Miller indices for a plane.

$$\sin^2 \theta_{hkl} = \frac{\lambda}{4a^2}(h^2 + k^2 + l^2)$$

The corresponding relation for the hexagonal system is

$$= \frac{\lambda^2}{4}\left[\frac{4h^2 + k^2 + kh}{3a^2}\right] + \frac{l^2}{c^2}$$

where 'd' is known 'a' can be calculated. By knowing a, l, and M, the number of atoms/molecules in the unit cell are calculated.

Generally the lines in diffraction pattern are enormous for non cubic systems and the lines are only few for cubic systems. Thus the line positions in the photograph of diffraction patterns indicates the shape and size of the unit cell. The line intensities in the photograph indicate the atom positions. For example let us consider NaCl crystal. When diffraction from (111) planes are examined the spectra of alternate orders are found to be weaker than the others. Therefore one can conclude that the alternate (111) planes are composed exclusively of either Na or Cl atoms. Since Cl atoms are heavier than Na atoms, they should scatter X-rays more strongly and hence those line intensities are stronger. Thus the unit cell of NaCl crystal is built with atoms of Na or Cl at its corners. In the case of cubic crystals, knowing the values of $(h^2 + k^2 + l^2)$ for different planes, one can easily identify the type of cubic crystals. If the individual values of h k l sets are purely odd or even, immediately one can identify that the given type of crystal belongs to FCC lattice. There are three main X-ray different methods by which the crystal structure can be analysed. The Laue method and the rotating method are for single crystals. The powder method is applicable to finely divided crystalline/poly crystalline specimen powder.

Laue method : A single crystal specimen is held stationary while irradiation with white X radiation. A photographic film is placed to receive the rays diffracted through the crystal or those diffracted in a back reflection direction. An interpretation of the reflection spots in the photographic film leads to know about the crystal structure of the specimen. When the grains of the crystal are very fine (0.4μm) then in the film continuous rings instead of spots are seen. When the grain size > 40μm, few number of spots are seen. Instead of spots, if the pattern contains streaks then the crystal is a strained one. It is difficult to use this method for the determination of unit cell dimensions because the exact value of λ is not known for the different reflections. The position of the Laue spot depends on the orientation of the crystal relative to the incident beam direction.

Powder method : This is also called Debye-Scherrer method. The powder is kept inside small capillary tube which would not undergo diffraction by X-rays. A narrow pencil of monochromatic X-rays is diffracted from the powder and recorded by the film as series of lines of varying curvature. The full opening angle of the diffraction cone 4θ is determined by measuring the distance S between two diffraction lines corresponding to a particular plane is related to the Bragg angle by the equation $4\theta = S/R$ radians, where R is the specimen to film distance, usually the radius of the camera housing the film. A list of θ values can be obtained from the measured values of S. Since the wavelength is known, substitution of θ and λ gives a list of spacing 'd'. Each spacing is the distance between neighboring planes (hkl). **Figure 3.1** presents the details of powder diffraction method. This method does not require

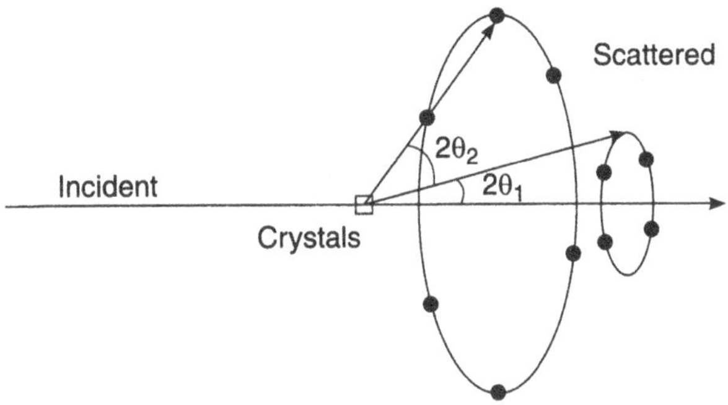

Fig. 3.1 Powder diffraction method.

single crystal. By measuring the distance between symmetrical lines (S) and the line intensities following information is obtained.

(a) A crystalline substance can be distinguished from an amorphous one because the former produce lines on a photograph the latter does not.

(b) If photographs are made before and after heat treatment, the comparison can reveal any alternation in the lattice.

(c) With the aid of standard photographs of known chemical compounds their presence may be detected in an unknown mixture.

(d) The size of the unit cell and the type of the lattice may be determined.

Rotating crystal method : The rotation makes it possible for the Bragg law to be satisfied. The reflections are usually recorded on a cylindrical film whose axis is the rotation axis of the crystal. The reflected spots lie of parallel lines. Reflections from planes indifferent orientations may overlap on the film and to make identification of the spots less, ambiguous, oscillation method and Weissenberg methods are used. They are used in complete crystal structure analysis.

Application : Evidently, nanoparticles are not very good single crystals and measurements in Laue geometry are not possible. One needs to perform powder diffraction experiments. Based on the combination of Bragg's Law and the properties of the 7 systems into which all crystal can be classified. One obtains equations which allow indexing of the observed XRD peaks. The next step consists in the determination of the number of atoms per unit cell. This is easily obtained, as the volume of the unit cell is readily calculated based on the parameters determined so far and usually the density of a given material is obtained easily and its chemical composition can be determined. The weight of a unit cell can be found and must correspond to an integer number of atoms of the respective types.

What remains to be done is the assignment of the atoms to the respective positions, which can be using the information contained in the intensity of the respective scattering peaks (after correcting for additional effects which can influence this intensity such as multiplicity, Lorentz absorption and temperature factors). Depending on the scattering power, absolute and relative intensities may vary. From the observations of how much this difference changes the intensity of a given peak can be derived which types of atoms are located at the lattice sites which contribute to a given Bragg peak.

In fact, this is the key problem of structural determination for a new class of material using XRD because a stepwise solution to the problem does not exist. Whereas in some cases comparison of the relative intensities of the respective Bragg reflexes and/or the information on the chemical composition can suggest some possible site occupations, more often than not a brute force approach has to be used distributing the atoms on the possible crystallographic sites and calculating the diffraction pattern which corresponds to this elementary cell of given symmetry and size. Naturally the number of feasible combinations scales with the size of the elementary and the number of atoms which it contains. This is also the reason why CPU power is a critical factor in protein crystallography. As far nanoparticles concerned, the resolving power of XRD is strongly correlated to the number of elementary cells in the crystallites which are investigated, because this exerts direct influence on the width of the diffraction lines. Broadening of the Bragg reflections is observed for particle with diameters < 100nm. It is possible to attempt to make use of the broadening e.g. by applying the Scherrer formula $D = 0.9\lambda/B \cos \theta_B$ in order to determine the particle size.

In most cases, the XRD data obtained on a given compound are compared to the huge crystal structure databases like the International Crystal Structure Database (ICSD) available today, which has opened up the possibility of fitting a given XRD spectrum by a method known as the Rietveld method. In this approach, a structural model is refined to reproduce the experimental data using intensity correction parameters, lattice parameters and possible zero shifts of the detection system, the line shape crystal structure parameters and the background. To achieve this fit after indexing the observed diffraction patterns and determination of possible space groups, lattice parameters are refined. Already during the process, the process of several phases can be identified. The rough structure obtained from this processing step is then compared to isomorphous compounds or compounds which feature a similar structure obtained in a database in order to obtain candidates for lattice site occupation. If the corresponding calculated diffraction pattern is similar to the observed one, refinement of the above parameters is performed. **Table 2** gives a Rietveld analysis.

TABLE 2. Rietveld analysis of 10 nm $Fe_{50}Co_{55}$ nanoparticle.

Relative	Contribution (%)	Error (%)
Amorphous	34.10	3.90
Corundum	11.23	1.11
α-iron	40.07	2.43
Maghemite	3.78	0.72
Magnetite	3.91	0.75
Quartz	4.23	0.93
Wuestite	2.65	0.57

Small Angle X-ray Scattering

In SAXS, elastic scattering processes into a given solid angle element are observed, but this time, the detector covers only small scattering angle (<1°). For a given X-ray, another ray with a path difference $\lambda/2$ also exists, if $\lambda \approx D \sin \theta$. For a given X-ray wavelength the determination of the threshold angle under which scattering can just be observed yields information on the dimension of the particle. To illustrate this, let us assume that we are dealing with homogeneous particles of uniform size and shape which are embedded in a homogeneous matrix. Local electron densities depends on the particle shape.

3.3 X-RAY ABSORPTION

X-ray absorption spectroscopy (XAS) measures the dependence of the cross-section of the absorption process on the energy of the incoming photon. Variation of photon energy in big steps over an extended energy range reflects the electronic structure of the corresponding element; steps in the cross-section occur whenever the photon energy is sufficiently high to excite electrons from a deeper core level; at the highest energy from the 1s level preceding with decreasing photon energy to 2s, $2p_{1/2}$, $2p_{3/2}$ and so on. However, there are bound unoccupied states and thus even at energies slightly lower than the ionization threshold it should be possible to exite the electrons into bound unoccupied states.

The absorption edge, i.e., the onset of the increase in the absorption cross section lies in fact at lower energies than the ionization energy. Usually at energies higher than the ionization threshold oscillatory structures are observed. This observation is explained in a simple model that the photo electron propagating through matter can be treated as a spherical wave, whose wave number K is connected the energy of the incoming photon E and the ionization energy $k = \sqrt{[2m/\hbar^2 (E-E_o)]}$ where, m represents the electron mass and $\hbar$ is the normalized Planck's constant. This photoelectron wave propagates through an environment in which it undergoes electron-electron scattering process. Due to interference of out going an scattered wave, what one sees is an interference pattern. On the other hand varying the energy of the incoming photon varies the wavelength of the photoelectron; the interference patterns changes towards destructive interference and back to constructive interference.

This explains the modulations usually seen in the absorption cross section. The scattering probability is a function of the energy of the photoelectron; consequently in the energy region just above the absorption edge multiple scattering will be possible and a large number of strong interference terms appear in the spectrum. These strong spectral features are called Shape Resonances. For historical reasons, the first two regions are known as X-ray absorption near edge structure (XANES) also called near edge X-ray absorption fine structure (NEAFS); latter is known as the extended X-ray absorption fine structure (EXAFS).

At the absorption edge, the unoccupied valence states are probed by the absorption process. As chemistry modifies the valence state/electronic structure of the elements so one obtains information on the chemical environment of the absorber atom in the sample which can be selected by choosing the excitation energy. At the same time the interference pattern is characteristic for a given arrangement of the atoms which surround the absorbing atom and allows the extraction of the information on its local coordination geometry without any need for long range order. If the onset of the structures shifts to higher energies with higher formal oxidation states, this rather systematic effect is called "chemical shift". Intensities of absorption in the white line range vary notably and this is correlated to the density of (atom projected) unoccupied states and tends to be higher oxidation states.

Applications : Cobalt nanoparticles are synthesized and stabilized by cetyl trimethyl ammonium bromide. Using XANES spectra, cobalt nanoparticles with sizes of 5.5, 8 and 11nm are analyzed. The intensity of the pre-edge structure of these spectra grows with the particle size where as the maximum absorption decreases. In general, resemblance to the hcp cobalt spectrum increases as the observed changes in intensity occur at those energy positions where spectral features in the cobalt foil spectrum are located but even the spectrum of the largest particles differs from pure hcp cobalt. This suggests that addition of the hcp of both the 11nm and the 8nm particles by linear super position of the spectra of the smallest particle whose diameter is 5.5nm and that of hcp cobalt. From this fit, an additional hcp content of ~ 40% for the 11nm particle and of ~18.5% for the 8nm particle can be derived. The error range for these numbers can be estimated to be ±5%. Assuming spherical particles and that the pre-edge intensity present in the spectrum of the smallest particle can be directly correlated to the amount of metallic cobalt present in this particle (40%), it is possible to extract the thickness of the respective cores and shells for the particles from the hcp cobalt contributions. The determined contribution of this phase is then given by the quotient of the cube of the core radius r_c and the cube of the total radius r_t. Performing this calculation, one obtains a core radius of 4.75nm for the particles of 11nm diameter, a core radius of 3.12nm for the particles of 8nm diameter and a core radius of 2nm for the 5.5nm particles. All values indicate that the shell thickness is always 0.8nm which in turn enhances the possibility of the assumptions.

The nature of the shell is identified. Likely candidates for the shell would be a cobalt oxide, such as CoO or Co_2O_3. The assumption of Co_2O_3 as being the compound forming the shell faces the problem that prolonged exposure to air converts the particles to CoO and annealing removes the shell which is inconsistent with assuming any type of oxide for the shell because of the high stability of oxides. The comparison of the spectrum of the smallest particle to the ones of the different Co (II) reference spectra indicates clearly that, under the assumption that the shell material dominates the constituents of the smallest particle, this material must contain Co (III). The shell is destroyed completely at very low temperatures of ~ 210°C more likely candidates for the shell are Co (III) complexes such as Co $[(NH_3)_6]Br_3$ and Co $[(NR_3)_6]Br_3$. The formation of these complexes and also the observed core shell structure of the particles seem to be "typical" of CTAB.

3.4 PARTICLE SIZE DETERMINATION

A method of determining the sizes of the particles is by measuring how they scatter light. The extent of scattering depends on the relationship between the particle size d and the wavelength λ of the light and it depends on the polarization of the incident beam. For example the scattering of white light which contains wavelengths in the range from 400nm for blue to 750nm for red, off the N_2 and O_2 molecules in the atmosphere with respective sizes d = 0.11 and 0.12nm explains why the light reflected from the sky during the day appears blue and that transmitted by the atmosphere at sunrise and sunrise appears red.

Particles size determinations are made using a monochromatic (single-wavelength) laser beam scattered at a particular angle (usually 90°) for parallel and perpendicular polarizations. The detected intensities can provide the particles size, the particle concentration and the index of refraction. The Raleigh Gans theory is used to interpret the data for particles with sizes $d < 0.1\lambda$, which corresponding to the beam nanoparticles measured by optical wavelengths. This method is applicable to those have diameters above 2nm and for smaller nanoparticles other methods are used.

The sizes of particles <2nm can be conveniently determined by the method of mass spectrometry. A gas mass spectrometer ionizes the nanoparticles to form positive ions by impact from electrons emitted by the heated filament (f) in the ionization chamber (I) of the ion source. The newly formed ions are accelerated through the potential drop in voltage V between the repeller (R) and accelerator (A) plates, then focused by lenses L and collimated by slits, s during their transit to the mass analyzer. The magnetic field B of the mass analyzer oriented normal to the page exerts the force $F = qvB$ which bends the ion beam through an angle of 90° at the radius r after which they are detected at the ion collector. The mass m, charge q ratio is $m/q = B^2 r^2 /2V$. The bending radius r is ordinarily fixed in a particular instrument, so either the magnetic field B or the accelerating voltage V can be scanned to focus the ions of various masses at the detector. The charge q of the nanosized ion is ordinarily known, so in practice it is the mass m that is determined. Generally the material forming the nanoparticle is known, so its density $\rho = m/v$ is also a known quantity and its size or linear dimension d can be estimated or evaluated as the $3\sqrt{V} = (m/\rho)^{1/3}$.

The mass spectrometer usually makes use of magnetic field mass analyzer. Modern mass spectrometers use other types of mass analysers such as the Quadrapole model or the time of flight type in which each ion acquires the same kinetic energy $1/2mv^2$ during its acceleration out of the ionization chamber so the lighter mass ions move faster and arrive at the detector before the heavier ions thereby providing a separation by mass.

3.5 STRUCTURE OF SURFACES

At low energies (10–100eV) the electrons penetrate only very short distances into the surface and their diffraction pattern reflects the atomic spacings in the surface layer. To know about surface layers of a material low energy electron diffraction (LED) is used, if the diffraction pattern reflects the atomic spacings in the surface layer. If the diffraction pattern arises from more than one surface layer, the contribution of lower-lying crystallographic planes will be weaker in intensity. The electron beam behaves like a wave and reflects from crystallographic planes in analogy with an X-ray beam and its wavelength λ, called the "de Broglie wavelength", depends on the energy E expressed in the units of electron volts through the expression $\lambda = 1.226/\sqrt{E}$ nm which is different from that of X-rays. Thus an electron energy of 25.2eV gives a de Broglie wavelength λ equal to the GaAs bond distance in GaAs ($\sqrt{3}a/4 = 0.2442$nm) where the lattice constant a = 0.565nm and the low energies are adequate for crystallographic electron diffraction measurements.

Reflection high energy electro diffraction (RHEED) is also used to determine surface layer lattice constants. This is carried out at grazing incidence angles where the surface penetration is small. When θ is small, λ must be small so the energy E must be large and hence the need for high energies for applying RHEED diffraction at grazing incidence.

3.6 SPECTROSCOPY

Atoms in a crystal are residing at particular lattice sites, but in reality they undergo continuous fluctuations in the neighbourhood of their regular positions in the lattice. These fluctuations arise from the heat or thermal energy in the lattice and become more pronounced at higher temperatures. Since the atoms are bound together by chemical bonds the movement of one atom about its site causes the neighbouring atoms to respond to this motion. The chemical bonds act like springs that stretch and compress repeatedly during oscillatory motion. The result is that many atoms vibrate in unison and this collective motion spreads throughout the crystal. Every type of lattice has its own characteristic modes or frequencies of vibration called "normal modes" and the overall collective vibrational motion of the lattice is a combination or super position of many, many normal modes. For a diatomic lattice such as GaAs, there are low frequency modes called acoustic modes, in which the heavy and light atoms tend to vibrate in phase.

A simple model for analyzing these vibratory modes is a linear chain of alternating atoms with a large mass M and a small mass m joined to each other by springs (~~) as follows: ~~~ m ~~ M ~~ m ~~ M. When one of the springs stretches or compresses by an amount Δx, a force is exerted on the adjacent masses with the magnitude $C\Delta x$, Where C is a spring constant. As the various springs stretch and compress in step with each other, longitudinal modes of each atom is along the string direction. Each such normal mode has a particular frequency ω and a wave vector $k = 2\pi/\lambda$ where λ is the wavelength and the energy E associated with the mode is given by $E = \hbar\omega$. There are also transverse normal modes in which the atoms vibrate back and forth in directions perpendicular to the line of atoms. For acoustic wave, increase of frequency ω increase wave number K while for optical wave it decreases in frequency. For the two waves the limiting frequencies are $(2c/M)^{1/2}$ and $(2c/m)^{1/2}$ with an energy gap between them at the edge of the Brillouin zone $k_{max} = \pi/a$ where a is the distance between atoms m and M at equilibrium.

If a one dimensional crystal has a lattice constant 'a' and a length that is $L = 10a$, then the atoms will be present along a line at positions $x = 0, a, 2a, 3a\ldots 10a = L$. The corresponding wave vector k will assume the values $k = 2\pi/L, 4\pi/L, 6\pi/L \ldots 20\pi/L = 2\pi/a$. The smallest value of k is $2\pi/L$ and the largest value is $2\pi/a$. The unit cell in this one dimensional coordinate space has the length 'a' and the important characteristic cell in reciprocal space is called the Brillouin zone and has the value $2\pi/a$. The electron sites within the Brillouin zone are at the reciprocal lattice points $k = 2\pi n/L$ where $n = 1,2,3\ldots 10$ and $k = 2\pi/a$ at the Brillouin zone boundary where $n = 10$.

Infrared and Raman Spectroscopy

The atoms in molecules undergo vibratory motion and a molecule containing N atoms has (3N-6) normal modes of vibration. Particular molecular groups such as hydroxyl-OH, amino-NH_2 and nitro NO_2 have characteristic normal modes that can be used to detect their presence in molecules and solids. The atomic vibrations correspond to standing wave types. This vibrational motion can also produce traveling waves in which localized regions of vibratory atomic motion travel through the lattice. Localized traveling waves of atomic vibrations in solids are called "Phonons" with energy $\hbar\omega = h\nu$.

In infrared spectroscopy and IR photon $h\nu$ is absorbed directly to induce a transition between two vibrational levels E_n and E_n' where $E_n = (n+1/2)\ h\nu_o$. The vibrational quantum number $n = 0,1,2,3\ldots$ is a positive integer and ν_o is the characteristic frequency for a particular normal mode. In accordance with the selection rule $\Delta n = \pm 1$, infrared transitions are observed only between adjacent vibrational energy levels and hence have the frequency ν_o. In Raman spectroscopy, a vibrational transition is induced when an incident optical photon of frequency $h\nu_{inc}$ is absorbed and another optical photon $h\nu_{emit}$ is emitted.

$$E_n = |h\nu_{inc} - h\nu_{emit}|$$

The frequency difference is given by $|\nu_{inc} - \nu_{emit}| = |n' - n''|\, \nu_o = \nu_o$. Since the same infrared selection rules $\Delta n = \pm 1$ is obeyed. Two cases are observed :

(1) $\nu_{inc} > \nu_{emit}$ corresponding to a stokes line,

(2) $\nu_{inc} < \nu_{emit}$ for an antistokes line.

Infrared active vibrational modes arise from a change in the electric dipole moment μ of the molecule, while Raman active vibrational modes involve a change in the polarizability $P = \mu_{ind}/E$ where the electric power vertor E of the incident light induces the dipole moment μ_{ind} in the sample. Thus some vibrational modes are IR active and some are Raman active.

IR and optical spectroscopy is often carried out by reflection and the measurements of nanostructures provide the reflectance R which is the fraction of reflected light. For normal incident $R = I_r / I_o = |\sqrt{\varepsilon}-1| / |\sqrt{\varepsilon}+1|$ where ε is the dimensionless dielectric constant of the material. The dielectric constant $\varepsilon(\nu)$ has real and imaginary parts, $\varepsilon = \varepsilon'(\nu) + \varepsilon''(\nu)$ where the real or dispersion part ε' provides the frequencies of IR bands and the imaginary part ε'' measures the energy absorption or loss. A technique called Kramers-Kronig analysis is used to obtain the frequency dependencies of $\varepsilon'(\nu)$ and $\varepsilon''(\nu)$ from the measured IR reflection spectra.

The classical way to carry out IR spectra is to scan the frequency of the incoming light to enable the detector to record changes in the light intensity for those frequencies at which the sample absorbs energy. A major disadvantage of this method is that the detector records meaningful information only when the scan is passing through absorption lines while most of the time is spent scanning between lines when the detector has nothing to record. To overcome this deficiency, modern infrared spectrometers irradiate the sample with a broad band of frequencies simultaneously and then carry out "Fourier transformation" to convert the detected signal back into the classical form of the spectrum. The resulting signal is called a Fourier transform IR (FTIR) spectra.

To learn more about an IR spectrum, the technique of isotropic substitution is employed. The frequency of a simple harmonic oscillator ω of mass m and spring constant c is proportional to $(c/m)^{1/2}$. So an isotropic substitution changes the IR absorption frequencies of chemical groups. The replacement of ordinary hydrogen ^1H by the heavier isotope deuterium ^2D (0.015% abundant) ordinary ^{12}C by ^{13}C, ^{14}N by ^{15}N and ^{16}O by ^{17}O. If deuterium is used for hydrogen, the absorption frequency shifts by $\sqrt{mD/mH} = \sqrt{2}$. For BN nano powder after deuteration, $\sqrt{2}$ shift is noticed. The deuteron converted the initial B-OH, B-NH_2, B_2-NH groups on the surface to B-OD, B-ND_2 and B_2-ND respectively.

The power of IR spectroscopy to elucidate surface features of nanomaterials is the study of $\gamma - Al_2O_3$. It has a defect spinel structure and its large oxygen atoms form a tetrahedral (VI) and two tetrahedral (IV) sites per oxygen atom in the lattice and Al^{3+} ions located at these sites are designed by the notations (VI) Al^{3+} and (IV) Al^{3+} respectively. There are a total of 5 configurations assumed by absorbed OH groups that bond to aluminium ions at the surface and these are shown in **Table 3**. The first two types $I_a = I_b$ and involve the simple cases of OH bonded to tetrahedrally and octahedrally coordinated Al^{3+} ions respectively. The remaining three cases involve the OH radical bound simultaneously to two or three adjacent trivalent Al^{3+} ions. The frequency shifts assigned to these five surfaces species are also seen by IR spectra.

Raman Spectroscopy

In this spectroscopy a phonon with a frequency of vibration in the IR region of the spectrum corresponding to ~400cm^{-1} or a frequency of 1.2×10^{13} HZ is used. Germanium nanocrystals prepared by chemical reduction, precipitation and subsequent annealing of the phase $Si_x\ Ge_y\ O_z$ when analysed by Raman spectra 20nm size particles are seen.

TABLE 3. Five possible configurations of adsorbed OH groups on γ–Al_2O_3 surface.

Frequency of OH^- (Cm^{-1})	Type	Configuration
>3750	I_a	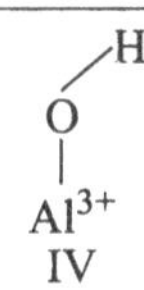
	I_b	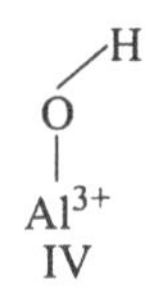
3720 to 3750	II_a	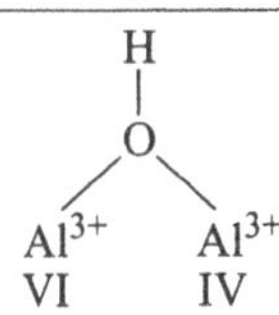
	II_b	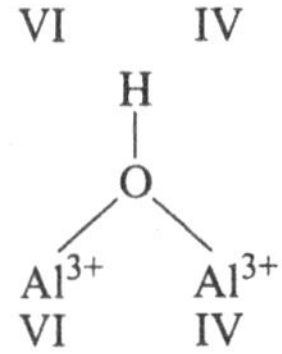
~3700	III	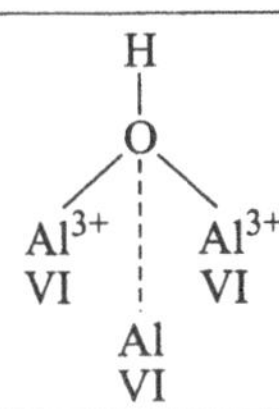

When a low frequency acoustic phonon is used, the process is called Brillouin scattering. Acoustic phonons can have frequencies of vibration or energies that are a factor of 1000 less than those of optical phonons. Brillouin spectroscopy involves both stokes and anti stokes lines. The optical mode corresponds to high frequency lattice vibrations and the acoustic mode conforms to lattice vibrations at much lower frequencies. The scattered frequency has the value $\omega_{scat} = \omega_{inc} \pm \omega_{phonon}$ where the negative sign in the expression for $\Delta\omega = \omega_{phonon}$ corresponds to a stokes line and the positivê sign denotes an anti stokes line. These two types of scattering that entail a change in frequency of the emitted photon are called inelastic. When there is no frequency shift ($\Delta\omega = 0$) then the scattering is the elastic Raleigh type that takes place in X-ray diffraction.

The Raman spectra of bulk crystalline Ge exhibits a narrow absorption line $\approx 3cm^{-1}$ wide arising from the $\Gamma_{25}{}^{+}$ optical phonon mode at the frequency of 300 cm^{-1}. Silicon nanoparticles exhibit the same behaviour as their Ge counter parts for the $\Gamma_{25}{}^{+}$ optical phonon mode and is centered at 521 cm^{-1}. The broadening and shift of the Si and Ge Raman lines to lower frequencies occur as the particle size decreases. This is due to photon confinement in nanocrystals.

Brillouin Spectroscopy

It is type of Raman scattering in which the difference frequency $\Delta\omega = \omega$ phonon corresponds to the acoustic branch of the phonon dispersion curves with frequencies in 10^9 HZ range. Brillouin scattering depends on particle size. For nano Ag particles deposited on SiO_2, the Raman shift and size are given as **Table 4.**

TABLE 4. Average particle size and Raman shift for μc-Ag in SiO_2.

Average Particle Size	Raman Shift (cm^{-1})
2.7	16.1
4.1	14.2
5.2	12.1

It was found that dependence of Raman shift was linear with (1/diameter) of the nanoparticles.

Photoemission Spectroscopy

Electron spectroscopic techniques are used to estimate quantitatively the elements present and the bonds between atoms. Bonds between individual atoms are caused by the overlap of atomic orbital and will individually approach the problem from the stand point of the formation of individual surface molecules. The orbital approach is used to under surface bonding. In addition to the valence electrons present, be it band like or molecular orbital like, each atom present (except H_2) possesses core electrons not involved in the bonding. The energies of these electrons are characteristic of the individual toms, yet they are also responsible (chemical shifts) to changes in the valence electronic structure of the system and so may be used to provide information on bonding.

In X-ray photo electron spectroscopy (XPS), an X-ray of known energy hν impinges on the sample, ejecting electrons from it with kinetic energies given by $E_n = h\nu - \beta E_n - \varphi$ where βE_n is the binding energies of the electrons with respect to the Fermi level and φ is the work function of the sample. The kinetic energies of the ejected electrons are measured in the electron spectrometer and the work function term refers to that of the spectrometer rather than the sample. This technique provides direct values for the core energy levels and for the energies of the electron in the valence bond. In vacuum UV photo electron spectroscopy (UPS) the photon energies used are insufficient to eject core electrons and one is limited to a study of the valence electrons. The instrumental advantages of using this technique for valence electron studies are the much lighter resolution obtainable (vacuum UV light source line widths are only a few meV where as soft X-ray line widths are only a meV) whereas soft X-ray line widths are only a 1eV and the higher intensity of the signal obtainable. The latter is ionization cross-section for the valence electrons with UV radiations.

In Auger electron spectroscopy (AES), the analysis involves secondary (Auger) electrons which may be ejected when a system with a core hole relaxes. The initial hole may be produced by several methods. Though the conventional method is by electron impact, Auger electrons will also be produced with photoelectrons in XPS. Since the energies of Auger electrons ejected in relaxation processes not involving the valence levels are determined by the energies of the core levels, one has an immediate atomic identification. In cases where valence levels, are involved in the relaxation process atomic identification is still usually possible as the core hole energy is usually the dominant energy term. The question of 'chemical shifts' in AES is more complex than for XPS. Since three energy levels involved, two of which may be valence. Each of these techniques have different strengths and weakness according to which aspects of sensitivity one is considering : (1) Escape depth, i.e., how thick a surface slice contributes to the signal; (2) Total signal strength, i.e., how quickly one can detect a signal; (3) Quantitative nature of the data; (4) Degree of chemical information and (5) Surface changes induced by the measuring technique.

The photo electron spectrometer involves X-ray photons incident on a specimen. The emitted photo electrons then pass through a velocity analyzer that allows electrons within only a very narrow range of velocities to move along trajectories that take them from the entrance slit on the left and allows them to pass through the exit slit on the right and impinge on the detector. The detector measures the number of emitted electrons having a given kinetic energy. The energy states of atoms or molecules ions in the valence band region have characteristic ionization energies that reflect perturbations by the surrounding lattice environment. So this environment is probed by the measurement.

An XPS study of 10nm InP provided the indium $3d_{5/2}$ asymmetric line. Analysis revealed the presence of two superposed lines; a main one at 444.6 eV arising from In in InP and a weaker one at 442.7 attributed to In in the oxide In_2O_3. The phosphorous 2p XPS spectrum exhibited two well resolved phosphorous peaks, one from InP and one from a P-O species.

NMR and EPR Spectroscopy

NMR spectroscopy involves the interaction of a nucleus possessing a none zero nuclear spin I with an applied magnetic field B_{app} to give energy level splitting into (2l+1) lines with energies $E_m = \hbar\nu\, B_{app}\, m$ where ν is the gyromagnetic ratio (magnetogyric ratio) characteristic of the nucleus and m assumes integer or half integer values in th range $-I < m < +I$ depending O whether I is an integer or a half integer. The value of n is sensitive to the local chemical environment of the nucleus and it is customary to report the chemical shift δ of n relative to a reference value ν_R, that is $\delta = (\nu - \nu_R)/\nu_R$. Chemical shifts are very small and are usually reported in ppm. The most favourable nuclei for study are those with I = ½ such as H, H_F, ^{31}p and ^{13}C.

Fullerene, C_{60} has the shape of a soccer ball with 12 regular pentagons and 20 hexagons. The fact that all of its carbon atoms are equivalent was determined unequivocally by the ^{13}C NMR spectrum that contains only a single narrow line. In contrast to this, the rugby-ball shaped C_{70} fullerene molecule which contains 12 pentagons (2 regular) and 25 hexagons, has 5 types of carbons and this confirmed by the ^{13}C NMR spectrum.

Electron Paramagnetic Resonance (EPR)

This is also called electron spin resonance (ESR). It detects unpaired electrons in transition ions, especially those with odd numbers of electrons such as Cu^{2+} (3d°) and Gd^{3+} ($4f^7$). The energies or resonant frequencies are three orders of magnitude higher than NMR for the same magnetic field. A different notation is employed for the energy, $E_m = g\, \mu_B\, B_{app}\, m$ where μ_B is the Bohr magnetron and g is the dimensionless g factor, which has the value 2.0023 for a free electron. For the unpaired electron with spin S = ½ on a free radical EPR measures the energy difference $\Delta E = E_{1/2} - E_{-1/2}$ between the levels m = ±1/2 to give a single line spectrum at the energy level $\Delta E = g\, \mu_B\, B_{app}$ and $g\, \mu_B = \hbar\nu$. If the unpaired electron interacts with a nuclear spin of magnitude I, then (2I+I) hyperfine structure lines appear at the energies. $\Delta E\,(m_I) = g\, \mu_B\, B_{app} + Am_I$ where A is the hyperfine coupling constant and m_I takes on the (2I+I).values in the range of $-I \le m_I \le I$.

EPR is used to study conduction electrons in metal nanoparticles; to detect the presence of conduction electrons in nanotubes and to distinguish metallic semiconducting properties. In the colloidal TiO_2 semiconductor nanoclusters, EPR is used to identify trapped oxygen holes. EPR is used in nanostructured biomaterials. Microwaves are used to study photon assisted SET and coulomb blockades in quantum dots.

3.7 LUMINESCENCE

The photoluminescence excitation technique involves scanning the frequency of the excitation signal and recording the emission within a very narrow spectral range. In bulk materials the luminescence spectrum often resembles a direct absorption spectrum. For 5.6nm CdSe quantum dot, the photoluminescence response appeared near 2.05eV. The excitation spectra of nanoparticle of CdSe with a diameter of 3.3nm exhibited the expected band edge emission at 2.176eV at 77K and they exhibited an emission signal a 1.65eV arising from the deep traps. A comparison of PLE spectra for the band edge and deep-trap emissions with the corresponding absorption spectra revealed the reflections the emission from only a small fraction of the overall particle size distribution. Shallow traps that can be responsible for band edge emission have the same particle size dependence in their spectral response. This considerably reduces the inhomogeneous broadening and the result is a narrowed nearly homogeneous spectrum. The emission originating from the deep trap does not exhibit the same narrowing. A shift of spectral line positions to higher energies occurs as the size of a nanoparticle decreases. CdSe nanoparticle exhibited a shift of the band edge gradually to higher energies when the size increased form 1.5nm to 4.3nm. The peak of fluorescence spectra shifted to higher energies as the excitation photon energy increased.

As nanoparticles get smaller and smaller, the percentage of atoms on the surface becomes an appreciable fraction of the total number of atoms. The fluorescent spectra of 3.4 nm and 4.3 nm CdS particles exhibited a sharp fluorescence at 435nm and 480nm respectively arising from excitons and a broad fluorescence emission at longer wavelengths. Addition of nitro methane quenched the fluorescence by bringing about a shift towards longer wavelengths plus an appreciable decrease in the magnitude of the broad band, and in addition it practically eliminates the sharp exciton emission. When the temperature raised from 4K to 259K, the peak exhibited a decrease in intensity and a shift towards longer wavelengths. This suggested that the hole traps be much deeper than do electron traps. The trapping of electrons in CdS nanoparticles is extremely fast (pico second) range requiring 10^{-13} s or less time to complete the trapping so all the trapped electrons are in place before there is an appreciable onset of the fluorescence.

Thermoluminescence

It is the emission by heating. In nanoparticles sometimes electron-hole pairs produced by irradiating a sample do not recombine rapidly but become trapped in separate metastable states with prolonged life times. The presence of traps is pronounced in small nanoparticles where a large percentage of atoms are in the surface, many with unsatisfied chemical bonds and unpaired electrons. Heating the sample excite lattice vibrations that can transfer kinetic energy to electrons and holes held at traps and there by release them with the accomplishment of emitted optical photons that constitute the thermal luminescence.

Gradually heating of the sample provides the energy needed to bring the release of electrons and holes from traps. The light emission as a function of temperature is recorded. The energy corresponding to the maximum emission is called the "glew peak". It is the energy needed to bring about the detrapping and is a measure of the depth of a trap. This energy however is generally insufficient to excite electrons from their ground states to excited states. At 300K, the thermal energy $k_BT = 25.85meV$ is far less than typical gap energies E_g, although it is comparable to the ionization energies of many donors and acceptors in semiconductors.

Zeolite-Y has a porous network of silicate and aluminate tetrahedral that form cube octahedral cages ~0.5 nm in diameter. As the CdS is introduced in the structure, the average size of the cluster increases. They exhibit a shift to red (toward longer wavelengths) as the loading increases from 1 to 5%. The smaller clusters associated with low loading have more surface states and hence more electrons to detrap and contribute to the glow peak. The increase in quantum confinement characteristic of smaller clusters also contributes to the increase in recombination probability with the resulting enhanced thermo luminescence.

3.8 MICROSCOPY

"Microscope" was first coined by members of the first "Academic dei Lincei", a scientific society which included Galileo. In 1590, Hans and Zacharias Janssen of Middle burg, Holland manufactured the first compound microscope. In 1660, Marcello Malpight observed capillaries. Robert Hooke in 1665 devised the compound microscope. He observed the thin slices of cork. He said, "I could exceeding plainly perceive it to be all perforated and porous .... These pores or cells .... were indeed the first microscopical pores I ever saw and perhaps, that were ever seen, for I had not met with any writer or person, that made any mention of them before this." In 1673 Antioni Van Leewen Hoek, Delft, Holland, created a "simple" microscope that could magnify to about 275 and published drawings of microorganisms in 1683. He discovered bacteria, free living and parasitic microscopic protests, sperm cells, blood cells and microscopic nematodes. He wrote letters to Royal society of London. George Adams Sr.(1740-1772) made microscopes. He made simple microscopes that could attain around 2μm resolution while the best compound microscopes were limited to 5μm because of chromatic aberration. In the 1730's, a barrister named Chester More Hall observed that the flint glass dispersed colours much more than "crown glass". He designed a system that used a concave lens next to a convex lens which could realign all the colours. This was the first achromatic lens. George Bass was the lens maker who actually made the lenses but he did not divulge the secret until over 20 years later to John Dolland who copied the idea in 1759 and patented the achromatic lens. In 1872 Glovanni Battista Amici built high quality microscopes and introduced the first matched achromatic microscope in 1872. In 1830, Joseph Jackson Lister solved the problem of spherical Aberration. In 1877, Ernst Abbe together with care Zeiss resolved distance of an objective. Abbe's law states that $d = \lambda/2n \sin\theta$ where d is the minimum resolving distance; λ = wavelength, θ = the angle of the light cone formulated glass lenses that colour corrected objectives and produced the first "apochromatic" objectives in 1886.

Early 20th century Professor Kohler developed a method of illumination called "Kohler illumination". This creates an evenly illuminated field of view while illuminating the specimen with a very wide cone of light. Two conjugate image planes are formed. One contains an image of the specimen and the other filament from the light direction change of a ray of light passing from one transparent medium to another with different optical density. Refraction is a ray from less to more dense medium is bent perpendicular to the surface with greater deviation for shorter wavelength. In diffraction light rays bend around edges. New wavefronts are generated at sharp edges. If the aperture is small, the definition is low. Dispersion is the separation of light into its constituent wavelength when entering a transport medium. An object can be focused generally no closer than 250 mm from the eye. This is normal viewing [1x magnification] young people may be able to focus as close as 125mm so they can magnify as much as 2x because the image covers a larger part of the retina.

Lenses focus light rays at a specific place called the focal point. The distance between the centre of lens and focal point is focal length. Simple microscopes have only a single lens. There are bright field microscope, dark field microscope, and phase contrast microscope and fluorescence microscopes. Modern microscopes magnify both in the objective and the ocular and thus are called "Compound microscopes". The bright field microscope produces a dark image against a brighter background. It has several objectives and the total magnification is the product of the magnifications of the ocular lens and the objectives lens. The dark field microscope produce a bright image of the object against a dark background. Phase contrast microscopy accentuates diffraction of the light that passes through a specimen.

Human eye can view objects of 0.02 cm. Ultramicroscope came with a resolution to observe 20Å unit dimension. The absolute theoretical limit of magnification of theoretical ultra microscope was about 3000 while in practice the magnifying power of 1000 alone was obtained. It is the light that sets the limit to magnification. The wavelength of visible light used for illumination in optical microscope is of the order of 0.6 nm, objectives smaller in dimension than this wavelength could not be seen since light would pass between them without being scattered by them.

Electron Microscopy

The electron microscopy is a device to magnify the minute objects where a beam of electrons is employed in stead of light rays as in the ordinary optical microscope. Earlier version was classified into two categories; the electrostatic which use the electrostatic lens system for focusing the electron beam and the magnetic where magnetic lens were used. They are further classified into emission type where the magnified image is simply a reproduction of the surface of the cathode, the source of electrons and the transmission type, where the magnified image is a reproduction of an object inserted between the cathode and the screen. **Table 5** compares the essential features of electron microscope and optical microscopes. **Principle of magnetic focusing :** Consider the motion of a charge in an uniform magnetic induction "B" . If the charge is initially moving in a plane normal to B, the force on it is also in this plane and normal to its direction of motion. Thus no work is done on the particle and its velocity remains constant in magnitude. But the charge will move in a circle in this plane and the force towards the centre of the circle is equal to $mv^2/r = evB$; The radius of the circle $r = (mv)/eB$; Hence angular velocity $\omega = Be/m$. This equation shows that ω and hence the time taken to make one resolution is independent of the velocity of the particle. This fact is used in magnetic focusing.

If the initial velocity of the charge is not normal to B, but makes an angle 'θ' with the direction of B, then one can resolve the velocity into a component vcosθ parallel to B and a component vsinθ normal to B. The vector product ($\vec{V} \times \vec{B}$) has no component parallel to B and the component of velocity vcosθ will continue unaltered. The projection of the motion on a plane normal to B will be a circle of radius r = (mvsinθ)/eB. But the actual path of the particle will be a helix due to addition of vcosθ component. One revolution of the helix is completed in a time 2πω=2πm/eB and the particle has then moved a distance OA = (2πmv cos θ)/eB in the direction of B. For small values of θ, this distance is independent of θ_m the first approximation (since cos θ = $1-1/2\theta^2$). Thus all electrons starting from O at small angles to OX are focused at A. This is the principle used in magnetic focusing. Hence if there are a number of electrons moving in different directions with different velocities, they will reach the point A at the same time. Thus the magnetic field acts on the electron beam in the same way as a lens acts on the beam of light. It is therefore called a Magnetic lens and its focal length is OA.

TABLE 5. Comparison between electron microscope and optical microscope.

Parameter	Electron Microscope	Optical Microscope
Source	Electron gun	Incandescent bulb
Condenser lens	Electromagnetic	Convex glass
Specimen	Thin metallic slide	Slide
Objective lens	Electromagnetic	Convex glass
Projector lens	Electromagnetic	Convex glass
Screen	Fluorescent screen	Ground glass
Medium	Vacuum	Air
Focal length	Variable	Cannot be varied
Magnification	10^6	10^3
Resolving power	2 A.U	2000 A.U

An electron microscope consists of an electron gun, condenser and projector lens. Vacuum is maintained inside to allow the passage of the electron beam. Electrons emitted by a hot W filament cathode are accelerated to form a high velocity beam by the anode. The accelerated electrons are condensed into a fine beam by a condensing arrangement called condensing magnetic lens. It may be a series of current carrying coils (producing magnetic fields). This condensed electron beam is made to pass through the object specially prepared in the form of a thin slice. Depending upon the density and thickness of the replica (specimen) at each point some of the electrons are absorbed or scattered while the remainder transmit. They are focused by the objective magnetic lens which focused and enlarges the electron beam and projector magnetic lens arranged one after, another, enlarges further and focuses the beam on a fluorescent screen. The function of the objective magnetic lens is similar to the objective lens in optical microscope and the projector magnetic lens is similar to that of an eyepiece.

Scanning Electron Microscopy

SEM is one of the most versatile instruments available for the examination and analysis of the micro structural characteristics of solids. When bulk objects are examined as high resolution as 5nm. **Figure 3.2** presents the schematic diagram of scanning electron microscope.

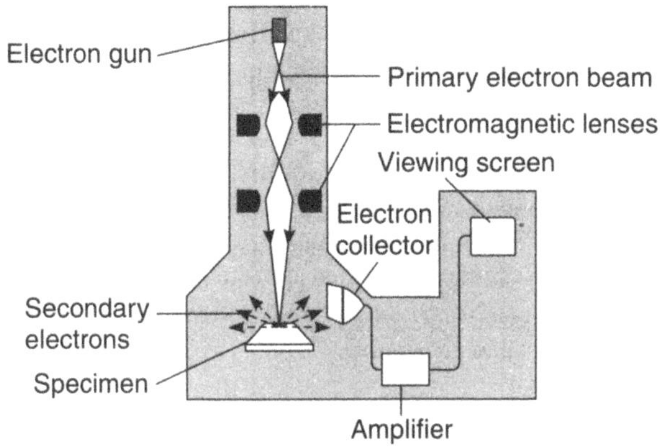

Fig. 3.2 Scanning electron microscope—schematic diagram.

The earliest work on the construction of a SEM is that of Von Ardenne in 1938. He added scan coils to a transmission electron microscope (TEM) and it was the first scanning transmission electron microscope. He obtained at a magnification of 8000 x, a spatial resolution of 50–100μm. In 1942, Zworykin and Co-workers examined thick specimens using SEM and obtained a resolution of ~50nm. In 1960 Everhart and Thomley (ET) increased the amount of signal collected and made an improvement in signal to noise (S/N) ratio. In 1963 Peak built a system known as SEM V with 3 magnetic lens, the gun at the bottom and E–T detector system.

Theory : The beam of electrons undergo a rich variety of interactions with the specimens. The interactions can be divided into two classes : (1) Elastic events which affect the trajectories of the beam electrons within the specimen without significantly altering the energy and (2) In elastic events which result in a transfer of energy to level leading to the generation of secondary electrons, Auger electrons, characteristics and continuum X-rays, long wavelength electromagnetic radiation in the visible, UV and IR regions, electron – hole pairs, lattice vibrations (phonons) and electron oscillation (plasmons). SEM and electron probe microanalysis (EPMA) are similar equipments. Manufactures make instruments capable of being operated both as SEM and EPMA. Usually the electron and X-rays optics are in combination, Both a secondary electron detector and an X-ray detector are placed below the final lens. The scanning unit allows both electron and X-ray signals to be measured and displayed on a cathode ray tube.

Electron guns : The electron beam current impinging on a specimen determines the source of the magnitude of the signals emitted. The gun provides a stable source of electrons which is used to form the electron beam. These electrons are obtained from a source of electrons which is used to form the electron beam. These electrons are obtained from a source by a process called "thermionic emission". At sufficiently high temperatures, a certain % of the electrons become sufficiently energetic to overcome the work function, Ew, of the cathode material and the escape the source. The cathode has a v-shaped tip which is about 5–100μm in radius. W or La B filament materials are used. Surrounding the filament is a grid cap or Wehnelt cylinder with a circular aperture centered at the filament apex. The grid cap is biased negatively between 0 and 2500 V with respectively through this voltage field and attempted to follow the maximum voltage gradient which is perpendicular to the field lines.

Electron lenses : The electron image formed by the beam at the point of cross oven is 10–50μM. The condenser and objective lens systems are used to demagnify this to final spot size of 5–200 nm. The condenser lens system which is composed of one or more lenses determines the beam current which impinges on the sample. The final probe forming lens often called the objective lens determines the final spot size of the electron beam and the current.

Electron probes : The minimum current sufficient to give a satisfactory scanning picture using secondary electrons is $\sim 10^{-12}$A. A 5nm electron beam can be obtained by W filament at 30KV and a 3nm electron beam using a La B filament at 30KV.

Image formation : The electron beam enters the specimen chamber and strikes the specimen at a single location. Within the interaction volume both elastic and inelastic scattering occur producing detectable signals form back scattered electrons, secondary electrons, absorbed electrons, characteristic and continuum X-rays and catho luminescence radiation. By measuring the magnitude of these signals with suitable detectors, a determination of certain composition etc., can be made at the single locations of the electrons beam impact. In order to study more than a single location the beam impact. In order to study more than a single location, the beam must be moved from place to place by the means of a scanning system. This is done as a function of time to sample the specimen properties at a controlled succession of points. Scanning is usually accompanied by a driving electromagnetic coils arranged in sets consisting of 2 pairs, one pair each for deflection in the x and y directions. In making a line scan the beam is scanned along a single line on the specimen e.g. in the x and y direction. The same signal which is derived from the scan generator is used to derive the horizontal scan of a cathode ray tube. The resulting synchronous line scan on the specimen and CRT produces a one to one correspondence between a series of points in the 'specimen space' and on the CRT or "display space". Thus the creation of an SEM image consists of constructing a map on the CRT. In SEM, image formation is produced by the mapping operation which transforms information from specimen space to CRT space. Magnification in the SEM image is accompanied by adjusting the scale of the map of the CRT. If information along a length of line l is the specimen space is mapped along a length L in the CRT space, the linear magnification is $M = L/l$. Taking only one or two images at a high magnification is not a satisfactory procedure as only a fraction of the area is sampled. So a combination of both low and high magnification is to be used.

Transmission Electron Microscopy

Electron beams not only are capable of providing crystallographic information about nanoparticle surfaces but also can be used to produce images of the surface and they play this role in electron microscope. In transmission electron microscopy (TEM) the electrons from a source such as electron gun enter the sample, scattered as they pass through it, focused by an objective lens are amplified by a magnifying (projector lens) and finally produce the desired image, in the manner reading from left to right. The wavelength of the electrons in the incident beam is $\lambda = (0.0388/\sqrt{v})$ nm, where the energy acquired by the electrons is E = eV and V is the accelerating voltage expressed in kilovolts.

If widely separated heavy atoms are present, they can dominate the scattering with average scattering angles θ given by $\theta \sim \lambda / d$ where d is the average atomic diameter. For an accelerating voltage of 100kV and an average atomic diameter of 0.15 nm we get $\theta = 0.026$ radius atoms interact with and absorb electrons to a different extent. When individual atoms of heavy elements are farther a part than several lattice parameters, they can sometimes be resolved by the TEM technique.

Electrons interact much more strongly with matter than do X-rays or neutrons with comparable energies or wavelengths. For ordinary elastic scattering of 100 KeV electrons the average distance traversed by electrons between scattering events called the "mean free path" varies from a few dozen nanometers for light elements to tens or perhaps hundreds of nm for heavier elements. The best results are obtained in electron microscopy by using film thicknesses that are comparable with the mean free path. Much thinner films exhibit too little scattering to provide useful images, and in thick films multiple scattering events dominate making the image blurred and difficult to interpret. Thick specimens can be studied by detecting back scattered electrons. Transmission electron microscope is shown as **Figure 3.3a.** Schematic diagram of transmission electron microscope is shown as **Figure 3.3b**. A TEM can form images by the use of the selected area electron diffraction (SAED) aperture located between the objective and projector lenses. The main part of the electron beam transmitted by the sample consists of electrons that have not undergone any scattering. The beam also contains electrons that have not undergone any scattering. The beam also contains electrons that have lost energy through in elastic scattering with no deviation of their paths and electrons that have been reflected by various hkl is inserted so that it allows only the main undeviated transmitted electron beam to pass.

The bright field image is observed at the detector or viewing screen. If the aperture is positioned to select only one of the beams reflected from a particular hkl plane, the result is the generation of a dark field image at the viewing screen. If the aperture is positioned to select only one of the beams reflected from a particular hkl plane, the result is the generation of a dark field image at the viewing screen. The details of the dark field image that is formed can depend on the particular diffracted beam (particular hkl plane) that is selected for the imaging.

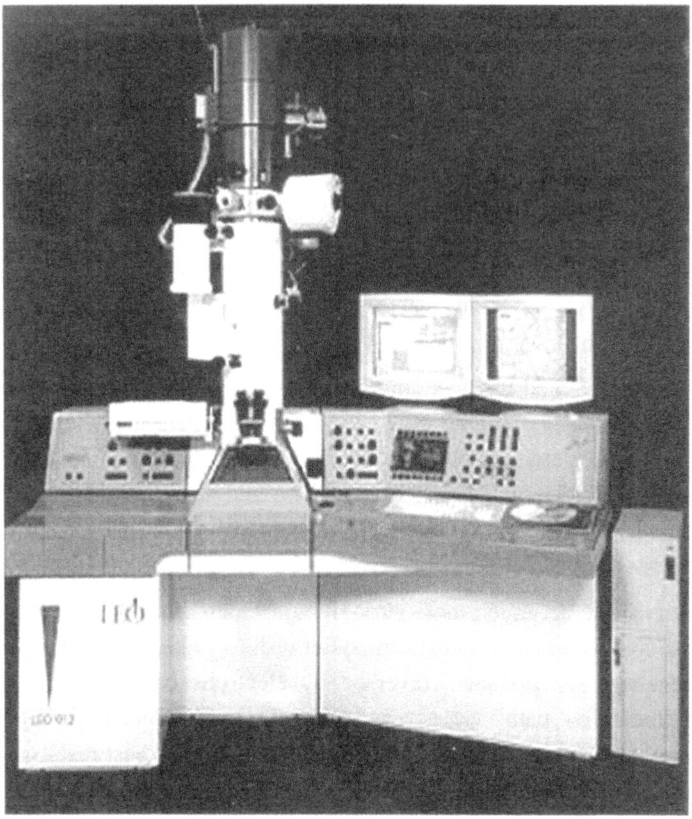

Fig. 3.3(a) Transmission electron microscope.

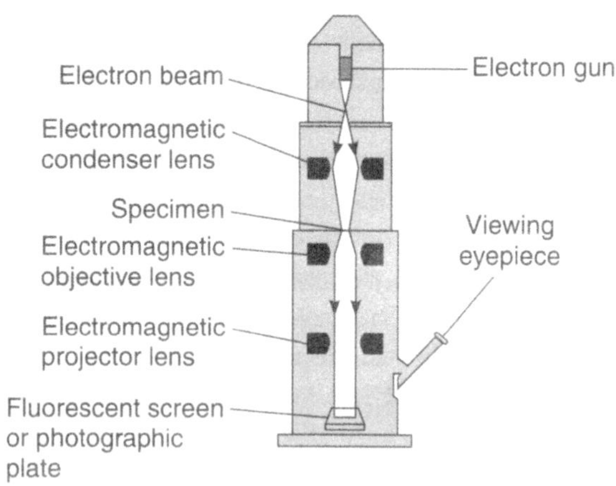

Fig. 3.3(b) Transmission electron microscope—schematic diagram.

A technique called image processing can be used to increase the information obtainable from a TEM image and enhance some features that are close to the noise level. If the image is Fourier transformed by a highly efficient technique called a "Fast Fourier transform" then it provides information similar to that in the direct diffraction pattern.

In addition to the directly transmitted and the diffracted electrons, there are other electrons in the beam that undergo inelastic scattering and lose energy by creating excitations in the specimen. This can occur by inducing vibrational motions in the atoms near their path and thereby creating phonons or Quantised lattice vibrations that will propagate through the crystal. If the sample is a metal then the incoming can scatter in elastically by producing a plasmon which is a collective excitation of the free electron gas in the conduction band. A third very important source of in elastic scattering occurs when the incoming electron induces a single electron excitation in the atom. This might involve inner core atomic levels of atoms, such as inducing a transition from a k level (n=1) or L level (n=2) of the atom to a higher energy that might be a discrete atomic level with a larger quantum number n, an electron band, or result in total removal (ionization) from the electron is in the valence band of a semiconductor. This excitation can decay via the return of excited electrons to their ground states there by producing secondary radiation and the nature of the secondary radiation can often give useful information about the sample. They can be surface sensitive because of the short penetration of distance of the electrons into the material.

Scanning Transmission Electron Microscopy

This is used for image analysis and for elemental micro analysis of areas as small as 50nm in size. In transmission electron microscopy, a beam of high energy electrons (100 – 400KeV) is collimated by magnetic lenses and passes through a specimen under high vacuum. The resulting diffraction pattern can be imaged on a fluorescent screen below the specimen. Form the diffraction pattern one can obtain the lattice spacings of the structure under consideration. Diffraction information from areas <0.1μm in size can be obtained. Alternatively one can use the transmitted beam or one of the diffracted beams to form a magnified image of the sample on the viewing screen. These are respectively the bright field and dark field imaging modes which give information about the size and shape of the micro structural constituents of the material with a resolution of ~0.2nm.

If the incident beam is allowed to raster and the transmitted beam is detected by a scintillater rather than by a photographic plate, a STEM image is obtained on a CRT. The signal from the CRT is coupled to a computer which allows the performance of various functions such as digital imaging image enhancement and particle size analysis. As the incident electron beam interacts with the specimen, characteristic X-rays are emitted by the sample. These can be detected and analysed. In this way, elemental composition from regions 0.05 μm in size can be obtained.

Scanning Tunneling Microscope (STM)

Since tunneling is a phenomenon known since the origin of quantum mechanics in the1920's, one may wonder why the tunneling microscope was not invented sooner. In 1930 there was an attempt to construct using air gap capacitors looking for the onset of tunneling. In 1986, Gerd Binning and Heini Rohrer obtained Nobel Prize in physics for the development of the STM. STM can operate not only in

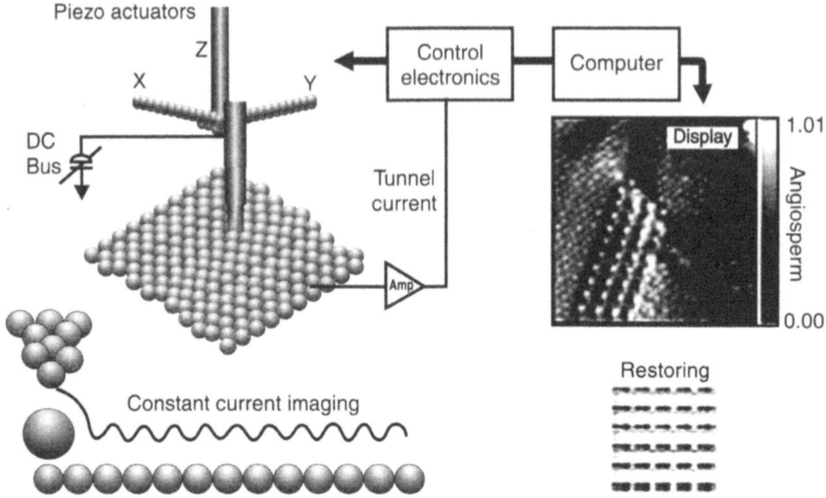

Fig. 3.4(a) Scanning tunneling microscope–schematic diagram.

Vacuum but also with the sample covered with electrolytes. Tunneling can occur as in vacuum. The components of scanning tunneling microscope are shown in **Figure 3.4a.** The tip as it travels on the surface of the sample is shown in **Figure 3.4b**. The inset shows the tunneling electrons.

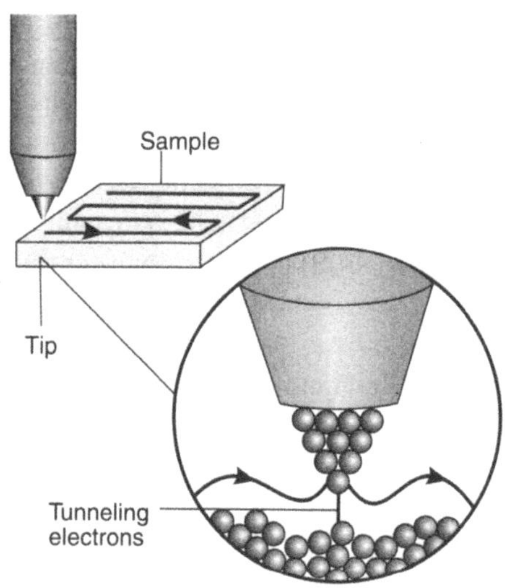

Fig. 3.4(b) Scanning tunneling microscope–tip and the sample.

A Scanning tunneling microscope utilizes a wire with a very fine point. This fine point is positively charged and acts as a probe when it is lowered to a distance of about 1nm above the surface under the study. Electrons at individual surface atoms are attracted to the positive charge of the probe wire and jump (tunnel) upto it there by generating a weak electric current. The probe wire is scanned back and forth across the surface in a raster pattern, in either a constant height mode or a constant current mode. In the constant current mode, a feed back loop maintains a constant probe height above the sample surface profile and the up/down probe variations are recorded. This mode of operation assumes a constant tunneling barrier across the surface. In the constant probe height mode, the tip is constantly changing its distance from the surface and this is reflected in variations of the recorded tunneling current as the probe scans. The feed back loop establishes the initial probe height and is then turned off during the scan. The scanning probe provides a mapping of the distribution of the atoms on the surface.

The SEM often employs a piezoelectric tripod scanner and an early design of the scanner was built by Binning and Rohrer. A piezoelectric is a material in which an applied voltage elicits a mechanical response and the reverse. Applied voltages induce piezo transducers to move the scanning probe (or the sample) in nm increments along the x, y and z directions. The initial setting is accomplished with the aid of a stepper motor and micrometer screws. The tunneling current which varies exponentially with the probe surface atom separation, depends on the nature of the probe tip and the composition of the sample surface. From a quantum mechanical point of view, the current depends on the dangling bond state of the tip apex atom and on the orbital states of the surface atoms.

Atomic Force Microscopy

AFM is widely used in nanostructures studies. The fundamental difference between the STM and the AFM is that the former monitors the electric tunneling current between the surface and the probe tip while the latter monitors the force exerted between the surface and the probe tip. The AFM can operate in a close contact mode in which the Core-to-Core repulsive forces with the surface dominate or in a greater separation "non contact" mode in which the relevant force is the gradient of the Vander Waals potential. As in the STM case, a piezoelectric Scanner is used.

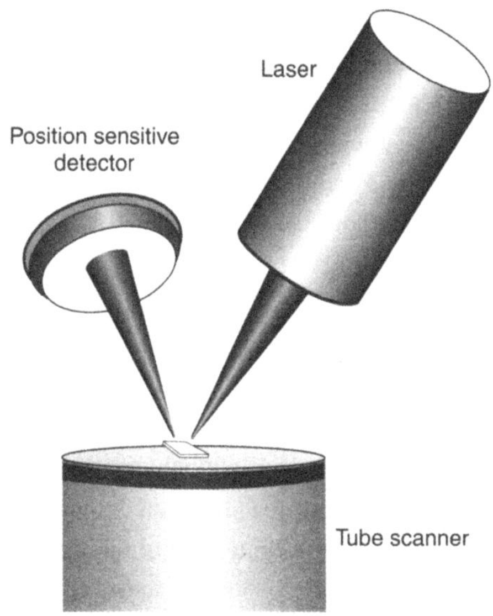

Fig. 3.5 Atomic force microscopy–schematic diagram

The vertical motions of the tip during the scanning may be monitored by the interference pattern of a light beam from an optical fiber or by the reflection of a laser beam. The AFM is sensitive to the vertical component of the surface forces. A related but more versatile device called a Friction Force microscope (lateral force microscope) simultaneously measures both normal and lateral forces of the surface of the tip. **Figure 3.5** shows a schematic diagram of AFM.

3.9 STANDARDS FOR NANOMETEROLOGY

Future nanotechnology needs measurement standards. Standard materials and meterology developed for use in nanoscale characterizations are the key tools for establishing reliability in information technology, biotechnology and environmental technology. The National meterology Institute of Japan (NMIJ) in the Institute of Advanced Industrial Science and Technology (AIST) is engaged in R & D for the nano meterology system ensuring the reliability of the product of national scientific and industrial activities under the auspices of the governmental nanotechnology program.

Three dimensional nanoscale certified Reference Materials Project is developing the nanometric 'scale' that can be applied to the technologies to control, process and measure the nanostructure with super fine precision. The 'nanoscale' is intended to offer the calibration traceable to the national measurements standards for lateral and depth directions and also to serve as the certified reference materials (CRMs) to which the certification value and uncertainty are assigned. NM1J / AIST is presently supplying the measurement standards for a one dimensional grading scale (0.2 to 8μm) and GaAl / AlAs super lattice reference material. The scanning electron microscope which is integrated with the calibrated standard microscale (240nm pitch) is marketed as the CD-SEM for semiconductors. In this project, the aim is to develop a nanoscale of 25–100nm pitch for the lateral direction, and of 3–10 nm thickness for the depth direction. These targets were decided by referring to the International Technology Roadmap for Semiconductors. (ITRS), which predicted the so called technology nodes by the half pitch of devices, would be 45nm and 22nm in 2010 and 2016 respectively.

Lateral Direction Nanoscale : The aim to develop a technology and calibration system for producing the nanoscale traceable to the length standards provided by the wavelength of an iodine-stabilized He Ne laser. It is also aimed at the scales will be supplied after calibration as CRMs in accordance with lateral nanoscales that have the shape of a one dimensional grating structure. One of the most promising methods for the calibration of nanometer length scales is AFM that has a resolution at the atomic level. A prototype system equipped with laser interferometers on the x, y and z axes has been developed. The uncertainty in measuring 240nm pitch is 0.1 7nm with 95% confidence. However in order to calibrate a scale with a minimum graduation of 25nm, it is required to make the overall un certainity even smaller. This is achieved by a calibration system (traceable AFM, J-AFM) which is an AFM equipped with laser inter ferometers having a resolution about 1/5th size of an atom and can be traceable to a length standard. It has:

(1) An xyz fine movement mechanism that is required for featuring angular variation during scanning along each optical axis to a few tens of nanoradians (nrad).

(2) The fine movement mechanism is mounted on a meterology frame featuring low thermal expansion to achieve the three-dimensional displacement with homodyne laser interferometers having a resolution of 0.02nm.

(3) An iodine stabilized offset lock laser is employed as the light source of the interferometers. This is traceable to the length standard and adopts a symmetrical optical configuration to reduce the effects of dead path.

(4) To reduce the cycle error, interferometers with uncertainty values below 0.1nm by building a system for reducing the cycle error. The broad band AFM head employs an AC mode cautilever and the AFM probe is in the form of a sharp-pointed probe.

The one dimensional diffraction grating that has size required as a CRM is fabricated by selecting the optimum nanofabrication technology from the technologies that have been established in the semiconductor industry such as superlattice film fabrication technology, electron beam lithography and X-ray lithography. Several kinds of lateral nanoscales with 100 to 25nm per graduation depending on the substances are under test by taking into consideration : (a) the user's convenience, (b) association with industry, and (c) the necessity of proving safety measures.

The nanoscales under development will have 10,000 to 40,000 graduations /1mm distance. Each of these many graduations contains small errors from the nominal value or variances due to imperfections in the fabrication process.

Depth Direction Nanoscale : A depth direction scale is required to quantify the properties such as the film thickness and the depth of injected impurities for example in the pn junction layer of the MOS field-effect transistor. Since most of the practical methods to measure the depth distribution depend on the material properties, unlike the in-plane direction nanoscales, the depth direction nanoscales necessitate the control of a wide range of factors besides the length (film thickness) such as the density of the films, the uniformity of their composition in a depth direction, the roughness of the surface and interface and the specific structures of the boundaries like a transition layer. GaAs/AlAs super lattices are fabricated up to 10nm level. Errors in measurements occur in SiO_2/Si ultra thin films due to structural transition layer due to oxygen when thermal oxidation is used. Ozone oxidation is being used. AIST established 100% concentration ozone gas and succeeded in low temperature fabrication of a high quality SiO_2 film on Si substrate. The thickness of the structural transition layer on the oxide film is extremely small and measured by chemical etching technique.

X-ray reflectivity technique (XRR) is a reliable method for thickness determination. When the angle of incidence of X-rays into a measurement sample exceeds a critical angle, their reflectivity suddenly decreases with increasing incidence angle and an oscillation structure appears, called the Kiessig Fringe. As the oscillation period is strongly related to the film thickness, the thickness can be determined by observing the oscillation structure while precisely controlling the incidence angle. To ensure traceability it is necessary to determine the X-ray incidence angle and the incident X-ray wavelength. Since the oscillation period increases as the film thickness decreases, the technique requires a high intensity X-ray source especially for the ultra thin films. For this purpose, the system uses an X-ray generator of 18KW out put from a rotating Cu target together with X-ray condensing optics. The scattering angle 2θ is measured accurately using a high resolution goniometer. The goniometer is controlled with an angle calibrator that is traceable to the national angle standard and the error in the angle measurement is reduced to below 1 arc second. This development would make it possible to implement a highly accurate film thickness calibration with an accuracy of less than one atomic layer.

XRR measurement for the GaAs/AlAs super lattice CRM is made and the least square fitting of the reflectivity vs 2θ revealed properties such as thickness, density surface roughness and interface roughness for all 4 layers. The repeatability of the thickness measurement was better than 0.5% except for the thickness of the top surface layer because it increased slightly with repeated measurements. The uncertainties are about 0.3nm with 95% confidence, the smallest among multi-layers known (Table 6).

TABLE 6. Evaluated properties of the GaAs/AlAs super lattice CRM.

	$\delta/10^{-6}$	$\beta/10^{-6}$	Thickness (nm)	Roughness (nm)
Oxide	8.132	0.266	1.241	0.361
GaAs	14.535	0.421	23.385	0.457
AlAs	10.709	0.296	22.572	0.334
GaAs	14.497	0.421	23.313	0.323
AlAs	10.581	0.296	22.589	0.361
Substrate	14.458	0.421	10000	0.349

Nano Process Technology and Nanometerology

Bureaue International des poids et Mesures (BIPM) compares various quantities in order to acquire an acceptable international uniformity. Various national meterology institutes (NMI) joined in the comparison and calibrated according to their own primary national length standards for nanometerology. The calibrations were made using optical diffraction (OD), optical microscopy (OM) and scanning probe methods (SPM). With regard to depth direction scales, the consultative committee for materials quantification (CCQM) has measured thickness of ultra-thin SiO_2 film or an Si substrate.

Mass/Diameter Measurement Technology

Particle measurement in gas phase : AIST has developed a method that enables highly accurate absolute measurements of mass for monodisperse particles suspended in air. The principle of this method is similar to that of the Milikan method, in that both work by balancing the electrostatic and gravitational forces experienced by charged particles suspended between two plate electrodes. The force balance is judged form the number of particles suspended after a certain hold tome. In this way, it can be applied to particles as small as 100nm, where as the conventional Millikan method would be unusable dice to Brownian motion of the particles. This new method is called the electro gravitational aerosol balance (EAB) method and combined with an accurate particle density determination in which particles are immersed in density reference liquids it gives accurate particles diameter. The EBA is now used to develop particle size standards for the particle size traceability system in Japan. AIST is trying to take EAB method one step forward so that is can be applied to even smaller particles. The instrument is aerosol particle mass analyzer (APM). It uses centrifugal force in stead of the gravitational force used in the EAB. It works as a continuous classifier of particles according to their mass to charge ratio. Combined with a condensation particle counter used downstream of the APM it can provide mass distribution of aerosol particles.

Particle measurement in liquid phase : Photon correlation spectroscopy functions on the principle that time correlation function of the light scattered from particles suspended in liquids is analyzed to determine the diffusion coefficient from which the particle diameter is derived. The diameters are < 100nm. The adoption of a dual correlator system, a high power YAG laser as the light source and a precise temperature control system has led to very accurate measurements. To determine particle size 1-20nm, nuclear magnetic resonance with pulled field gradients is being tried.

Nanopore measurements : AIST is developing a compact and easy to use position lifetime spectrometer for use in small laboratories and will take high-sensitivity nanoporosimetry. The positron implanted into an insulator such as silica pairs with an electron to form positronium. Positronium annihilates after a short life time, the duration of which depends on the size of the nanopores. The nanopore size increases from 0.5 to 2.5 nm with additive concentration in the Precursor solution.

Surface structure : XPS offers a measurement and analysis technique for use in non destructive and depth profiling surface analyses of nm order.

Thermal properties : AIST has developed a thermal diffusivity measurement technology picosecond thermo reflectance method. This detects surface temperature variations in the reflectance ratios that depend on temperature by heating the boundary between the thin film and the transparent substrate using a picosecond pulse laser (one picosecond is one-trillionth of a second). The technique is used measure the thermal diffusivity of single-layer thin films with thicknesses < 1μM and one of the boundary thermal resistances of multilayer thin films. As the measurement targets have been limited to metallic films with thicknesses ~100nm this sub theme aims at developing a technology that can measure the thermal diffusivity and thermal resistance of non metallic thin film boundaries.

With the former picosecond thermo reflectance method, drift and output variation in the light intensity from the laser had been reflected directly in the amplitude signal there by leading to deterioration of the signal to noise ratio. Meterology institute of Japan invented a detection method that is free from the effects fluctuation in heat source light and this is observing the phase component instead of the conventionally observed amplitude component. This has drastically improved the signal quality. Highly reliable measurements of thermal diffusivity of in metallic thin films having a thickness of ~100nm. Another problem with this previous picosecond thermo reflectance method is the restriction of the observable time span of measurements to about 1ns after the picosecond pulse heating. This restriction is due to adjustment of the parallelism of the optical axis of the optical delay path. This is overcome by extending the time span for observation of transient temperature changes to more than ions by oscillating a pair of picosecond titanium-sapphire lasers synchronously and controlling the time interval of oscillation electrically. This method allowed to measure the thermal diffusivity and thin film boundary thermal resistances of non-metallic thin films.

Low expansion glasses and as thermal glasses are important functional materials in the precision equipment, semiconductor and in aerospace. At nanometric resolution, development of a technology for measuring the thermal expansion coefficients of solid materials is in progress in AIST.

❑❑❑

CHAPTER 4

Applications of Nanotechnology

4.1 NANOBIOLOGY

Nanotechnology has been a domain of physics, chemistry, electrical engineering and material science. Macroscopic world properties do not vary considerably from the microscopic world, but at the nanometer scale (<300nm) materials can all together behave differently. This is because of interfacial forces and quantum size effects. Many biological materials are nanoparticles. Bacteria, which range in size between 1 and 10μm are in the mesoscopic range while viruses with dimensions from 10-200nm are at the upper part of the nanoparticle range. Proteins which ordinarily come in sizes between 4 and 50nm are in the low nm range. Proteins are amino acids tied together by peptide chemical bonds and form long polypeptides. The polypeptide nanowires undergo twisting and turnings to compact themselves into a relatively small volume corresponding to a polypeptide nanoparticle with a diameter that is typically in the range of 4–50nm. Thus a protein is a nanoparticle consisting of a compacted polypeptide nanowire. The genctic material deoxyribo nucleic acid (DNA) also has the structure of a compacted nanowire. Its building blocks are 4 nucleotide molecules that bind together in a long double helix nanowire to form chromosomes which in humans contain about 140×10^6 nucleotides in sequence. Thus the DNA molecule is a double nanowire, two nucleotide nanowires twisted around each with a repeat unit every 3.4nm and a diameter of 2nm.

Human tendon attaches muscle to a bone. The fundamental building block of a tendon is the assemblage of amino acids (0.6nm) that form the gelatin like protein called collagen (1nm) which coils into a triple helix (2nm). There follows a three fold sequence of fiber like or fibrillar nanostructures, a microfibril (3.5nm), a sub fibril (10-20nm) and a fibril itself (50-500nm). The final two steps in the build up, the cluster of fibers called a fascicle (50-300μm) and the tendon itself (10-50cm). Since the smallest amino acid, glycine is ~0.42nm in size and some viruses reach 200nm.

Nanobio-structures : If the crystal structure is known for the building block molecule and there are n-molecules in the crystallographic unit cell, then one can divide the unit cell volume V_u by n and take the cube root of the result to obtain an average size or average dimension. $d = (V_u/n)^{1/3}$. One can deduce the size by reconstructing the molecule from knowledge of its atomic constitution taking into account the lengths and angles of the chemical bonds between its atoms.

Another common way is to determine the size of a biological molecule is to observe using an electron micrograph. Many biological macromolecules such as proteins are characterized by their molecular weight M_W and the size of the nano particle is related as $d = 0.1184\ (M_W/\rho)^{1/3}$ nm) where ρ = density. Usually the densities are in the range of 1.43 to 1.607. The twisting and turnings of polypeptide nanowires to form a compact between structure held together by weak hydrogen and disulphide bonds can be somewhat loose with spaces present between the polypeptide nano wire sections. Hence the density of the protein is less than that of its constituent amino acids. Using electron microscopy, density can be calculated as $\rho = 0.001661$ M_W in the Daltons and V in cubic nm. **Table 1** summarizes the biological substances in the nm range. Amino acids have –COOH group, $-NH_2$ group and a group R that characterizes the acid are shown as **Table 2.**

Polypeptide nanowires and protein nanoparticles : Aminoacids combine together in chains through the formation of a peptide bond. To form this bond the –OH of the carboxyl group of one amino acid combines with the hydrogen atom H of the amino group of the next amino acid with the establishment of a C–N peptide bond accompanied by the release of water. Small peptides are called oligo peptides and amino acids incorporated into polypeptide chains are often referred to as "amino acid residues" to distinguish them free or unbound amino acids. The protein haemoglobin contains 4 polypeptides each with about 300 amino acid residues.

The stretched-out polypeptide chain is called the primary structure. To become more compact locally the chains either coil up in a what is called an alpha helix (α helix) or they combine in sheets called beta sheets (β sheets) held together by hydrogen bonds. The sheets might also be called Nanofilms. These two configurations constitute what is referred to as the secondary structure. An overall compactness is achieved by a tertiary structure consisting of a series of twisting and turnings held in place by disulphide bonds. If there is more than one polypeptide present, they position themselves relative to each other in a quaternary structure packing is a characteristic of globe shaped proteins. Some proteins are globular and others are elongated. In practice some of the space within a protein molecule residing in the cytoplasm of a cell will contain water of hydration between the twisting and turnings.

TABLE 1. Various biological substances in nm range.

Class	Material	M_W (Da)	Size d (nm)
Amino acids	Glycine	75	0.42
	Tryptophan	246	0.67
Nucleotides	Cytosine monophosphate	309	0.81
	Guanine monophosphate	361	0.86
	Adenosine triphosphate	499	0.95

Common molecules	Steric acid	284	0.87
	Chlorophyll	720	1.1
Proteins	Insulin	6K	2.2
	Haemoglobin	68K	7.0
	Albumin	69K	9.0
	Elastin	72K	5.0
	Fibrinogen	400K	50
	Lipoprotein	1300K	20
	Ribosome		30
	Glycogen		150
Viruses	Influenza		60
	Tobacco mosaic		120
	Bacteriophage T_2		140

TABLE 2. Molecular weight and size of few amino acids.

Name	M_w (D_a) Size d, nm	RNA words (CODONS)
Glycine	75.07 Da 0.42 nm	GGU GGA GGC GGG
Alanine	89.09 Da 0.47 nm	GCU GCA GCC GCG
Valine	117.12 Da 0.54 nm	GUU GUA GUC GUG
Threonine	119.12 Da 0.54 nm	ACU ACA ACC ACG
Glutamate	147.13 Da 0.7 nm	GAA GAG
Tryptophan	246.27 Da 0.9 nm	UGG

Nucleic acids : DNA is a nucleotide which is more complex than an amino acid. It contains a five membered deoxyribose sugar ring in the centre with a phosphate group (PO_4H_2), attached at the other end. The acid-phosphate group and the sugar group parts of adjacent nucleotides bond together to form the sugar-phosphate backbone of a DNA strand, resulting in a macroscopically long double stranded molecule. The complementary base pairs are held together between the two strands by H bonds. Weak hydrogen bonds are used to accomplish this. So the double helix can easily unwind for the purposes of transcription (forming RNA) or replication. The individual strand is 0.34nm thick, the double helix has a diameter of 2nm and the repeat unit containing 10 nucleotide pairs is 3.4nm long. To accomplish this coupling together of the two nanostrands in an efficient manner, a small single ring pyrimidine base always pairs off with a larger two ring purine base namely cytosine with guanine and thymine with adenine. The 2nm wide strands are many orders of magnitude too long to fit lengthwise in the nucleus of a 6μm diameter human cell so they undergo several stages of coiling. The next coiling stage consists of an approximate 140 base pair length of DNA winding around a group of proteins called histones to form what is sometimes called a "bead", which has a diameter of 11nm. The histone beads are joined together by lengths of double stranded DNA in the "linker region" between the heads. The DNA associated with the histones is called "chromatin" and the histone bead with encircling DNA strands is called a nucleosome. The linker regions provide the nucleosome sequence with the great flexibility that is required for stages of folding **Table 3.** The basic building block of DNA is a nucleotide and is more complex than an amino acid. It contains a five membered deoxyribose sugar ring in the centre with a phosphate group (PO_4H_2) attached at one end and a nucleic acid base R attached at the other end. On a DNA strand, the acid phosphate group and the sugar group parts of adjacent nucleotides bond together to form the sugar phosphate backbone of a DNA strand resulting in a macroscopically long double stranded molecule. Weak hydrogen bond are used to accomplish this so the double helix can easily unwind for the purposes of transcription (forming RNA) or replication (duplicating itself). The undividual strand is 0.34 nm thick, the double helix has a diameter of 2nm and the repeat unit containing 10 nucleotide pairs is 3.4 nm long. The 2nm wide strands are many orders of magnitude too long to fit lengthwise in the nucleus of a 6 micrometer dia meter human cell so they undergo several stages of coiling. **Figure 4.1** presents the structure of DNA molecule.

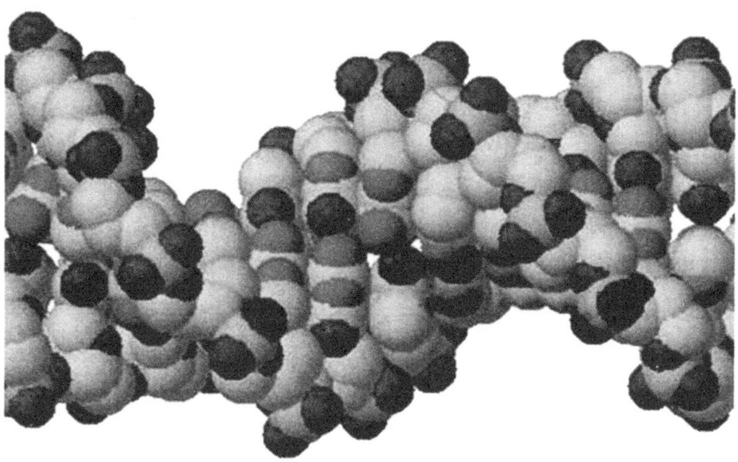

Fig. 4.1 DNA molecule.

In the next stage of compaction the nucleosomes stack one above the other alternating between two coiled columns and connected by the linker strands. They now form a structure of 1mm long chromatin fibers 30nm in diameter a configuration called the "L" packing of the nucleosomes. The chromatin fibers undergo the next higher order of folding of the 300nm wide foldings. These various stages of compaction are held in place largely by relatively weak hydrogen bonds which makes it feasible for the overall structure to unfold, partially or completely for replication during cell division or for transcription during the formation of RNA molecules which bring about/direct the synthesis of proteins. The final condensed structure of the metaphase chromosome is small enough to fit inside the nucleus of a cell. Figure 4.1 presents the structure of DNA.The genome is the complete set of hereditary units called genes, each of which is responsible for a particular structure or function in the body. In human beings the genome consists of 46 chromosomes with an average of ~1600 genes arranged lengthwise along each chromosome.

TABLE 3. Successive twisting and folding during the packing of DNA.

DNA	Size (nm)
DNA double helix	2
Chromatin	11
Packed nucleosomes	30
Extended section	300
Condensed section	700
Metaphase	1400

Genetic code : The DNA molecule contains the information for directing the synthesis of proteins. One of the DNA strands called the coding strand stores this information while the other strand of the pair, called the Complementary strand does not have useful data. The information for the synthesis is contained in a sequence of 3 letter words using a 4 letter alphabet A,C,G and T based on the 4 nucleic acid bases adenine (A), cytosine (C), guanine (G) and thymine (T). As there are 3 letters in a word and 4 letters in the alphabet there are $4^3 = 64$ possible words and 61 of these are used as code words for amino acids.

Proteins : A segment of the DNA molecule uncoils and the region of the double helix that stores the code words for that particular protein serves as a template for the synthesis of a single stranded messenger RNA molecule (mRNA) containing these code words. In transcribing the code, each nucleotide base of RNA is replaced by thymine in the RNA. Since DNA inventories the code words for the thousands of proteins and mRNA contains the code words for one protein, the mRNA molecule is very short compared to DNA, it is a short nanowire. The mRNA brings the message of the amino acid sequence from the nucleus of the cell where the transcription from DNA takes place to nanoparticles called Ribosomes in the cytoplasm region outside the nucleus where the proteins are synthesized. A number of additional protein nanoparticles called Enzymes function as catalysts for the protein synthesis.

Micelles and vesicles : Surfactants readily adsorb at an oil-water or air-water interface and decrease the surface tension there. It is a molecule characterized by a dimensionless packing parameter p defined as $p=(V_T/A_H L_T)$ where A_H is the area of the polar head and V_T and L_T are the volume and the length of the hydrocarbon tail. If the average cross sectional area of the tail (V_T/L_T) is less than that of the head $(p<1/3)$ then the tails will pack conveniently inside the surface of a sphere enclosing oil with a radius $r<L_T$ suspended in an aqueous solution. Such a structure is called a micelle. The elongated or cylinder-shaped micelle appears over the range $1/3<p<1/2$. Larger fractional values of the packing parameter $1/2<p<1$ lead to the formation of the vesicles which have a double layer surface structure. For example sodium di-2-ethyl hexyl phosphate can form nanosized vesicles with $V_T \sim 0.5nm^3$ $L_T >> 0.9nm$ and $A_H \sim 0.7nm^2$ corresponding to $p \sim 0.8$ which is in the vesicle range. If the packing parameter is one, $p = 1$; then the average transverse cross sectional area AH of the planer interface or they can form a bilayer. The inverse of a spherical micelle called an 'inverse micelle' occurs with $p>1$ for surfactants at the surface of a spherical drop of water in oil.

An emulsion is a cloudy colloidal system of μm size droplets of one immiscible liquid dispersed in another. If surfactants are present, then nanosized particles ~100nm can spontaneously form as a stable, transparent microemulsion that persists for a long time. These molecules in a solvent can organize in various ways depending on their concentration. For low concentration, they can adsorb at an air–water interface. Above a certain surfactant concentration, called critical. Micellar concentration, distributions of micelles in the size (2–10nm) constituted and disassembled with lifetime measured in μs or seconds.

If the vesicles are formed nutural or synthetic polypholipids, they are called unilaminar or only one bilayer. A phospholipid is a lipid substance containing phosphorous in the form of H_3PO_4 which functions as a structural component of a membrane. The hydration causes them to spontaneously self assemble into unilaminar liposomes. Unilaminar liposomes have diameters from the nm to the μM ranges with bilayers that are 5–10nm thick.

Biomimetrics involves the study of synthetic structures that mimic or imitate structures found in biological systems. It makes use of large scale or super molecular self assembly to build up hierarchial structures similar to those found in Nature. The process of biomeralisation involves the incorporation of inorganic compounds such as those containing calcium into soft living for the self assembly of many these films involved in biomineralization can be modeled as dimenzation $R + R \rightleftharpoons R_2$ with a low equilibrium constant $K_D = C_D/(C_F)^2$; $R_n + R \rightleftharpoons R_{n+1} \ldots$ with a much larger equilibrium constant $K = C_{n+1} / C_F C_n$. These two constants exercise control over the rate at which the reaction proceeds for the case under consideration, $K_D < K$ the concentration of free or inbound monomers C_F always remains below a critical concentration $C_o = 1/k$ specially $C_F < C_o$. When the total concentration of free and clustered monomers C_T satisfies the condition $C_T < C_o$, then the free monomer concentration C_F increases with C_T. When a high enough concentration is provided so that C_T becomes larger than the critical value (i.e., $C_T > C_o$) then the aggregate forms and grows for further increases in C_T. In analogy with this model, self assembly kinetics involves a slow dimer formation step follows by faster propagation steps with $K_D << K$.

Bio-assemblies : Assembly is the arrangement of atoms in a desired fashion with greater precision and flexibility. There are two categories viz; the self assembly and the positional assembly. In self assembly the parts move randomly under the influence of thermal noise and possible mutual orientations are explored until a stable arrangement is obtained. Positional and self assemblies are interlinked. All of the assemblies lie between these two types on a scale depending on positional uncertainty when the positional uncertainty is high, it is more of a positional assembly.

The requirements of the building blocks in the above two cases are not the same. In positional assembly, the building blocks must not only link to each other, but they must also bind and release from the positional device. Secondly the linkage groups between the blocks in the positional assembly are very strong and rigid as compared to the multiple weak bonds of self assembly. In the self assembly, these linkages provide a high degree of specificity in inter building block reactions. In positional assembly building blocks should not allowed to encounter with each other in uncontrolled random orientations.

Proton pumps : These are molecular structures, possessing the feature of producing continuous renewable conversion of light to chemical, mechanical or electrical energy. The process of synthesizing ATP requires energy. The proton gradients across the cell membrane created by enzymes help ATP synthase (ATPase) synthesize ATP. When protons circulate through these devices they provide the means of coupling the exchange of free energy. When an ATPase reverses its role, it helps to develop a proton gradient and is called a pump.

PS is a proton source which initiates the proton transfer and is located near the centre of bilayer of the membrane. PS transfers the proton to the primary proton acceptor A1 which could be a part of the side chain of the amino acids in the protein. From A1 it is transferred to A2 and so on towards the cell membrane until it is captured by a water molecule and is carried to the aqueous environment. The key feature required in this mechanism is to inhibit back proton transfer form A1 to PS. This could be achieved by making the proton transfer between A1 and A2 fast and practically irreversible.

One potential application of proton pumps is drug delivery using pH sensitive liposome systems which are stable at neutral pH sensitive liposome systems which are stable at neutral pH and release their contents on acidification. This release could be triggered by tissue or skin illumination if the liposome membrane is contained in proton pumps.

Dendrimers : The ubiquitous structure is a highly branched pattern occurring at all levels of the biological hierarchy, is endowed with the unique interfacial and functional performance. The patterns are found at all levels of dimension scales. These structures are also called class IV polymers. The surface groups rise exponentially on increasing the diameter linearly. Thus "tethered congestion" occurs to produce geometrically closed structures, exhibiting guest host container properties.

There are two major strategies dendrimer synthesis. The first one is the divergent method in which growth of the dendron starts from the core. The other method proceeds from the molecular surface to the core. Proteins have a tertiary structure and are fragile but contrastingly dendrimers are 3 dimensional robust structures covalently fixed which are also hollow in the interiors (nanoscale container). Dendrimers also exhibit the property of nano scaffolding clustering together in an exo-receptor with a wide variety of biological molecules.

Biological detection and imaging : Quantum dots are non-scale crystalline structures which can transform the colour of light. The quantum dot is considered to have greater flexibility than other fluorescent materials. Colloidal nanoparticles have been linked with biomolecules such as protein and peptides. These bio conjugates provide the key to assembling new materials developing homogeneous bioassays and as multicolour fluorescent labels for ultra sensitive detection and imaging. An important aspect of quantum dots is their photo stability which allows the intracellular processes to be monitored for a longer period of time. Multiple targets inside living cells can be targeted with the help of multicolour nanocrystals. Quantum dots are applied for in vivo molecular imaging with the help of near IR fluorescent nanocrystals. A further application is in multiplexed optical encoding and high throughout analysis of genes and proteins by various colours and intensities of quantum dots.

4.2 NANOCATALYSIS

Catalysis can play two principle roles in nanoscience. Catalysis can be involved in some methods for the preparation of Quantum dots, nanotubes and a variety of other nanostructures. Some nanostructures can serve as catalysis. Nanoparticles have an appreciable fraction of their atoms at the surface. Their activity depends on surface area. However the catalytic activity will not necessarily scale with the surface area in the nanoparticle range of sizes. The activity of or turnover frequency (TOF) of the cyclohexene to cyclohexane normalised to the concentration of surface Rh metal atoms decreases with particles size from 1.5 to 3.5nm and then begins to level off.

The specific surface area gram S is $A/\rho V$ where ρ is the density (g /cc). A sphere of diameter d has the area $A = \pi d^2$ and the volume $V = \pi d^3/6$ to give $A/v = 6/d$. A cylinder of diameter d and length L has the volume $V = \frac{\pi d^2 L}{4}$. The limit $L << d$ corresponds to the shape of a disk with the area $A \sim pd^2/2$ including both sides to give $(A/v \sim 2/d)$, in like manner, a long cylinder or wire of diameter d and length $L >> d$ has $A \sim 2\pi rL$ and $A/V \sim 4/d$. For various geometries, using m^2/g as units

$$S(r) = 6 \times 10^3/\rho d \text{ for sphere of diameter d}$$

$$S(r) = 6 \times 10^3/\rho a \text{ for cube side a}$$

$$S(r) \sim 2 \times 10^3/\rho L \text{ thin disc} <<< d$$

$$S(r) \sim 4 \times 10^3/\rho d \text{ long cylinder or wire } d << L$$

Where the length parameters a,d & L are expressed in nm and the density ρ has the units g/cc. For GaAs, $\rho = 5.32$ g/cc and using this value specific surface areas of GaAs spheres long cylinders and thin disks as a function of size can be calculated (**Table 4).** For a cube of side a with the same volume as a sphere of radius r, $[4\pi r^3/3] = a^3$ so $a = [(4\pi/3)^{1/3} r]$. $S_{cub} = 1.24\ S_{sphere}$ and a cube has 24% more specific surface than a sphere with the same volume.

TABLE 4. Specific surface areas of GaAs objects.

Size (nm)	Surface area (m^2/g)		
	Sphere	Wire	Disk
4	281	187	94
6	187	125	62
10	112	76	37
20	57	38	19
30	38	26	13
40	29	19	10
60	19	13	6
100	11	8	4
200	6	4	2

If a cylinder of diameter D and length L with same volume as a sphere of radius r specifically $(4\pi r^3/3) = \pi D^2L/4$ considered the $r = 1/2\,[3D^2L/2]^{1/3}$. It is shown that specific surface areas (L/D)

$$S(L/D) = \left[\frac{\pi DL + 1/2\,\pi D^2}{\rho\pi D^2 L/4}\right]$$

Figure 4.2 presents an arrangement of a nanoparticle in a catalyst. For example gold nanoparticles supported on TiO_2 substrates show high activity for oxidation of Co at room temperature. Geometric effects lead to higher activity and selectivity for certain reactions. For example reaction $CH_2 = CH–CN + H_2O \rightarrow CH_2 = CH–CONH_2$ on Pd–Cu nanoparticles produced when 3:1 Cu:Pd ratio was used.

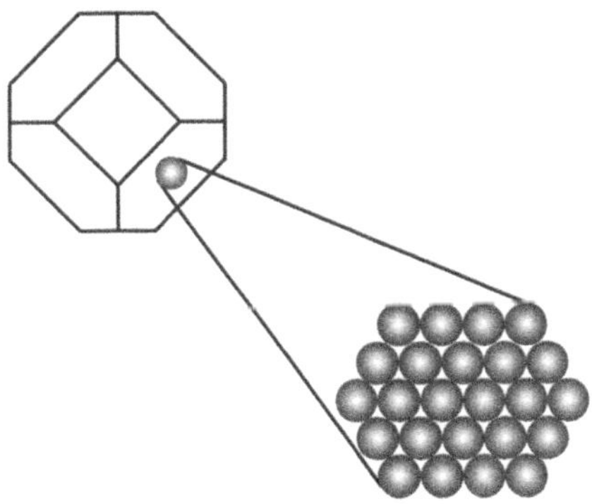

Fig. 4.2 Nanoparticle in a catalyst.

Molecular sieves : A molecular sieve which is a material suitable for filtering out molecules of particular sizes ordinarily has a controlled narrow range of pore diameters. Silicas and aluminas which can be prepared so that they have a porous structure of a more or less random type; that is they serve as sponges on a mesoscopic or μM scale. It is quite common for these materials to have pores with diameters in the nm range. Pore surface areas are sometimes determined by the Brunauer-Emmett-Jeller (BET) adsorption isotherm method in which measurements are made of the uptake of a gas such as N_2 by the pores.

Organo silicate molecular sieve MCM – 41 has a mean pore diameter of 3.94 nm. The introduction of relatively large trimethyl silyl groups $(CH_3)_3Si$ to replace protons of silanols SiH_3OH in the pores occludes the pore volume and increases the distribution of pores to a smaller range of sizes.

Pillared clays : Clays are layered minerals with spaces between the layers where they can adsorb water molecules or positive and negative ions and undergo exchange interactions of these ions with the outside. An idealized clay structure with the prototype talc formula $Mg_3Si_4O_{10}(OH)_2$. In an idealized structure it has 4 oxygen plains of the silicate layer which is perfectly flat with 2/3 of the oxygens in the 2 centrally located planes replaced by (OH) groups. All the tetrahedral sites are occupied by Si and all the octahedral sites are occupied by Mg forming the coordination groups, SiO_4 and MgO_6.

Saponite is a clay belongs to most morilbonite class. In this some aluminium (Al^{3+}) replaces silicon in tetrahedral sites and some divalent iron (Fe^{2+}) replaces. Mg in octahedral sites corresponding to the typical formula. $[Na_{0.3}Ca_{0.15}]$ $[Mg_{2.9}Fe^{2+}_{0.1}]$ and $[Si_{3.6}Al_{0.4}]$ $O_{10}(OH)_{2}.4H_2O$. Each layer is 0.94nm thick with a top and bottom surface area of 660m^2/g. It is possible to intercalate or place between the layers bulky ions that from pillars that hold the layers apart and there by provide a system of spaces or pores where various small molecules can reside. The dimensions of the pores produced by this layering process are in the low nm range. These materials are called pillared inorganic layered compounds (PILCs).

The pillaring is often carried out with the aid of the positively charged Al13, Keggin ion, which has the formula $[Al_{13}O_4\ (OH)_{24}\ (H_2O)_{12}]^+$ or alternatively $\{AlO_4[Al(OH)_2H_2O]_{12}\}^{7+}$. In this ion the centrally located aluminium is tetrahedrally coordinated and the surrounding 12 aluminiums are octahedrally coordinated. The Keggin ion is sometimes prepared in $AlCl_3$ solutions by transforming pairs of the trivalent hexa aquo ions $Al(H_2O_6)^{3+}$ to dimmers $Al_2(OH)_2(H_2O)_8{}^{4+}$, which are subsequently coalesced to form the Keggin ion. The presence of this ion can be unambiguously established by an ^{27}Al NMR measurement as the centrally coordinated aluminum ion produces a narrow NMR line that is chemically shifted by 6.2 ppm relative to hexa aquo aluminum.

One way to carry out the pillaring process is to suspend the clay in a solution containing Keggin ions at a controlled pH and a controlled OH : Al ratio. Some of the initial charge of +7 on the ion is compensated for when they bond with the silicate layers to form the pillars. The pillars themselves may be considered as cylinders with a diameter of 1.1nm. There are about 6.5 saponite unit with the oxygen content $O_{20}\ (OH)_4$ per pillar. Taking the value A = 3.15nm^2 for the area of the silicate layers in 6.5 unit cells the distance between nearest-neighbour pillars can be estimated. If the pillars are assumed to form a square lattice, then the spacing between the center points of nearest-neighbour pillars is $(A)^{1/2}$ = 1.77 nm and if they are arranged on a regular triangular or hexagonal lattice, then their separation is $(2A/\sqrt{3})^{1/2}$ = 1.90nm. Taking an average of these two values one gets a free space of ~0.74 nm between pillars. Experimental measurements indicate that the resulting pillared clay has a basal. Spacing of >> 1.85nm, a surface area of >> 250m^2/gm and a pore volume of >> 0.2 cm^3/g.

There are other types of montmorillonite clay materials also serve as pillars. Non alumina pillars are based on the proposed Zirconium ion $Zr_{18}\ O_4\ (OH)_{36}\ (SO_4)_{14}$, the trivalent titanium ion $[Ti(CH_3COO)_{6.4}(OH)_{0.4}Cl_{1.2}]Cl.11H_2O$, hexavalent chromium octahedral forming the ions $[Cr_4(OH)_6(H_2O)_{11}]^{6+}$ and $[Cr_4O(OH)_6(H_2O)_{10}]^{5+}$, an alumina silica. $Al_2O_3.SiO_2$ combination and SiO_2 supplemented by some titania TiO_2, they provide catalysts with a wide range of pore sizes.

4.3 NANOELECTRODES

In the 1960's ultra microelectrodes were discovered which have higher mass transfer efficiency, smaller RC cell time constants, lower IR drop, higher signal to noise radius varies form μm to nm.

Nanoimprint technology : The technology has three steps in which a mold with nanostructures on its surface is pressed into a thin resist cast on a substrate, followed by the removal of the mold. This step duplicates the nanostructures on the mold in the resist film. In other words, the imprint step creates a thickness contrast pattern in the rest. The second step is the pattern transfer where an anisotropic etching process such as reactive ion etching (RIE) is used to remove the residual resist in the compressed area. This step transfers the thickness contrast pattern into the entire resist. The lithography technique can pattern nanostructure over a large area with high through put and low cost.

Nanoporous SAM : A self assembled monolayer (SAM) based electrode was fabricated on a gold surface. This uses a hydrophilic 4 aminothio phenol (4-ATP) and a hydrophobic organothiol, n-alkane thiols. In the mixed SAM islands of electro active 4-ATP aggregates formed among electrochemically inert n-alkanethiols and the SAM'S acted as a micro electrodes array.

Template synthesis : This is a general method of preparing nanoelectrodes. It involves the synthesis of the nanostructure of a desired material with in the pores of a nanopores membrane. Since the channels with uniform diameter, mono disperse nanowires of the desired material are obtained. Depending on the chemistry of the pore wall and material, these nanowire may be either hollow (tubule) or solid (a fibre).

Etch coating : There are two processes. First metal wires or carbon fiber are etched to sharp tips by electrochemical method or flame etching. Second an insolated layer is coated with the tips. By increasing the temperature, the insulated layer contracts and solidified and the shape tip of the metal wires or carbon fibers is exposed with the desired diameter. Pt nano electrodes, Pt-Ir nanoelectrodes and carbon fiber nanoelectrodes have been fabricated using this method.

Nanomaterials coated electrode : Carbon nanotubes dispersed in Nafion DMF, DHP to give homogeneous dispersions and form stable films on electrode via solvent evaporation. Carbon nanotubes–ARS composite film on carbon electrode via electro deposition can be made.

Electrochemical properties : If the nanoelectrode is used in a potential step experiment, for spherical diffusion using Laplace transformation Fick's second law of diffusion becomes.

$$i(t) = [nFADC^{*}/r] + [nFA\sqrt{D}C^{*}/\sqrt{\pi t}]$$

where n = is the number of electrons transferred in the redox reaction, F is Faraday's constant A is the geometric electrode area, r is the distance from the centre of the sphere, C* and D are the concentration and diffusion coefficient for the electroactive species respectively. The above equation has a time-independent and a time-dependent part. The part which dominates depends on whether the experimental timescale is short or long. When the time scale is short, the diffusional layer thickness is smaller than the electrode radius and the spherical electrode appears to be planar to a molecule at the edge of this diffusion layer. The mass transport is dominated by semi infinite linear diffusion. The current at nanoelectrode follows the Cottrell equation $i(t) = nFA\sqrt{D}C^{*}/\sqrt{\pi t}$. At long time scales, the diffusional-layer is larger than the electrode radius and the mass transport is dominated by spherical diffusion. The current attains a time independent steady value as follows :

$$I = nFAC^{*}/r$$

The analysis on the parameter $\frac{(\pi Dt)^{1/2}}{r}$ can give the time scale at which the steady state current will dominate the total current. In nanoelectrodes the steady state current is obtained at very short time scale. D/r, the mass transfer coefficient, gives a very high radial diffusion rate at the nanoelectrode. Thus convection in solution does not cause obvious interference on the current. This makes nanoelectrodes appropriate for application in flow systems.

Nanoelectrodes array (NEA) eliminates the low single to noise ratio. It helps to detect low currents. This increases high mass sensitivity, increased mass transport and a decreased influence of the solution resistance, size reduction of each individual electrode to nm and an increase in the total number of electrodes can improve the detection limits and the signal to noise ratio since the noise level depends on the active area of the individual electrode and the signal depends on the total area of the electrodes. On NEA, the mass transport depends on the distance between individual nanoelectrodes diffusion time interval and the potential scan rates. There are 3 types of mass transport occurs.

1. Under radial diffusional conditions prevailing at the individual nano electrode with no overlap with the neighbouring elements, the steady state current of NEA's is the total i_{SS} of each nanoelectrode. This condition occurs when electroactive species are transferred at the very beginning or at very high scan rates or the density of NEA's is very low. At low scan rates or with the increased mass transfer radial diffusion boundary layers totally overlap inducing the NEA's to behave as a planar electrode.

2. Under planar electrode conditions, the signal to noise improves several amplitudes since the noise level depends on the active area of the individual electrode where as the signal depends on the total area of the electrodes.

3. At intermediate scan rates (250-3000mV/sec) the radial diffusion at neighbouring nanoelectrode begin to overlap and the non linear and radial diffusion are mixed.

The cell time constant and IR drop: A double layer is formed when an electrode is dipped in a solution. If the applied potential is changed, a charging current flows to the double layer capacitance. In a potential step experiment, the amplitude ΔE, the charging current IC decreases exponentially with the time at a rate observed by the cell time constant RC where R is resistance. For a nanoelctrode RC, cell time constant is very small and practically there is no charging current. The nanoelectrodes are suitable to study fast reactions. The IR drop depends on the resistance of the electrode and the solution resistance between working and reference electrodes. R increases as the radius of the electrode decreases. But on nanoelectrodes, the current flow is small (10^{-9} to 10^{-12} A), the product of IR is small. One can apply a two electrode configuration when using a nanoelectrode in electroanalysis, with the concentration of electrolyte very low or even zero. This characteristic enables nanoelectrodes to be utilized in high resistance or electrolyte media.

Characterization of nanoelectrodes : All the characterization methods employed to study nanoparticles are useful STM, AFM, SEM are popular. The active area can be calculated by steady state voltammetry and the effective radius of nano electrode is $I = 2\pi n \, FDC^* r$.

Industrial Applications

1. The detection of single molecules, their characterization and their chemical and physical manipulation are possible with their advances. Enzymes are immobilized on Pt-Ir wire of a nanoelectrode and electron transfer processes are studied. DNA molecules are electrostatically trapped between two metal nanoelectrodes separated by 8nm and the current flow through the DNA is measured.

2. Needle type Pt disk nanoelectrodes as extremely miniatured scanning probes for high resolution SECM.

3. A multiwalled carbon nanotube modified glassy carbon electrode is used to determine trace amounts of cadmium and lead.

4. Carbon fiber nanoelectrodes (200–300nm) is used to monitor the dopamine release of single Pc12 cells. A W/WO_3 nanosensor is used to measure the extra cellular pH variation of endothelial cells and also the pH variation of the normal damaged and recovery endothelial cells. Multiwalled carbon nanotube arrays have the characteristics of recording and stimulation of the brain via implanted nanodevices. Nanoelectrode arrays are to be used for epileptic focus detection and seizure abortion by controlled focal electrical stimulation.

5. Glucose biosensors based on carbon nanotubes are used. Nanoelectrodes are used for DNA detection. A multiwalled carbon nanotubes film coated carbon fiber nanoelectrode is used to monitor the NO release from liver mitochondria in "carassius auratus"

4.4 NANOMACHINES—ENGINEERING

Silicon integrated circuits produced machines and circuits having micrometer dimensions. A combination of lithography and metal deposition produced micro electron mechanical systems (MEMS). These devises are miniatures and multiplicand. Multiplicity refers to the large number of devices and designs that can be rapidly manufactured. Miniaturization enabled development of microsized devices. In MEMS, as the ratio of the surface area to the volume of a component is much larger than in conventional sized devices, this makes friction more important than inertia. In the micro world mechanical behaviour can be altered by a thin coating of a material on the surface of a component. Electrostatic forces become large. Electrostatic actuation is often used in micro machines which means that the elements are charged and the repulsive electrostatic force between the elements causes them to move.

Nanomotors exist in biological systems such as the flagellar motor of bacteria. Flagelle are long thin, blade like structures that extend from the bacteria through water. These whip like structures are made to move by a biological nanomotor consisting a highly structured conglomerate of protein molecules anchored in the membrane of the bacterium. The motor has a shaft and a structure about the shaft resembling an armature. However the motor is not driven by electromagnetic forces but rather by the breakdown of adenosine tri phosphate (ATP) energy rich molecules, which causes a charge in the shape of the molecules. Applying ratchet enables the protein shaft to rotate.

Electron beam and X-ray lithography can be used to make nanostructures. Electro-beam lithography uses a finely focused beam of electrons which is scanned in a specific patterns over the surface of material. It can produce a patterned structure on a surface having 10nm resolution. Because it requires the beam to hit the surface point by point in a serial manner, it cannot produce structures at sufficiently high rates to be used in assembly line manufacturing process but its mask technology and exposure are complex.

Nano-imprint lithography patterns have a resist by physically deforming the resist shape with a mold having a nanostructure pattern on it rather than be modifying the resist surface by radiation as in conventional lithography. A resist is a coating material that is sufficiently soft that an impression can be made on it by a harder material.

A mold having a nanoscale structured pattern on it is pressed into a thin resist coating on a substrate creating a contrast pattern in the resist. After the mold is lifted off, an etching process is used to remove the remaining resist material in the compressed regions. The resist is a thermo plastic polymer, which is a material that softens on heating. It is heated during the molding process to soften the polymer relative to the mold. The polymer is generally heated above its glass transition temperature there by allowing it to flow and conform to the mold pattern. The mold can be a metal, insulator or semiconductor. Fabricated by conventional lithography can produce patterns on a surface having a 10nm resolution at low cost and high rates of production.

STM has been used to build nanosized structures atom by atom on the surface of materials. An adsorbed atom is held on the surface by chemical bonds with the atoms of the surface. When such an atom is imaged in an STM, the tip and the adsorbed atom to the surface and the adsorbed atom will not be disturbed by the passage of the tip over it. If the tip is moved closer to the adsorbed atom such that the interaction of the tip and the atom is greater than that between the atom and the surface, the atom can be dragged along the tip. At any point in the scan the atom can be reattached to the surface by increasing the separation between the tip and the surface. In this way adsorbed atoms can be rearranged on the surface of materials and structures can be built on the surfaces atom by atom. The surface of the material has to be cooled to liquid He temperatures in order to reduce thermal vibrations, which may cause the atoms to diffuse thermally thereby distorting the arrangement of atoms being assembled. Thermal diffusion is a problem because this method of construction can be carried out only on materials in which the lateral or in-plane interaction between the adsorbed atom and the atoms of the surface is not excessive. The manipulation also has to be done under ultra. High-vacuum conditions in order to keep the surface of the material clean.

STM is used to manipulate a circular array of iron atoms on a copper surface. **Figure 4.3** depicts a circular array of iron atoms on a copper surface, called a "quantum corral" assembled by STM manipulation. The wave like structure inside

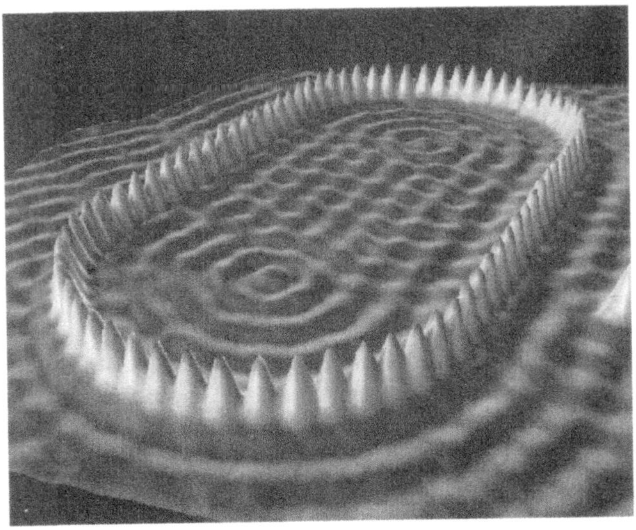

Fig. 4.3 Quantum corral

the corral is the surface electron density distribution inside the well corresponding to 3 quantum states of this two dimensional circular potential well, in effect providing a visual affirmation of the electron density predicted by quantum theory. This image is taken using an STM with the tip at such a separation that it does not move any of the atoms. The adsorbed atoms in this structure are not bonded to each other. The atoms will have to assembled in 3 dimensional arrays and be bonded to each other to use this technique to build nanostructures. Because the building of 3 dimensional structures has not been achieved, the slowness of this technique together with the need for liquid cooling and high vacuum all indicate that STM manipulation is a long way from becoming a large scale fabrication technique for nanostructures. It is important, however in that it demonstrates that building nanostructures atom by atom is feasible and it can be used to build interesting structures such as the quantum corral in order to study their physics.

The wave like structure inside the corral is the surface electron density distribution inside the well corresponding to 3 quantum states of this two dimensional circular potential well, in effect providing a visual affirmation of the electron density predicted by quantum theory. This image is taken using an STM with the tip at such a separation that it does not move any of the atoms. The adsorbed atoms will have to be assembled in 3 dimensional arrays and be bonded to each other to use this technique to build nanostructures. STM manipulation is not possible in a larger scale.

4.5 NANOWORKSHOP

Single walled carbon nanotubes deform when they are electrically charged. An actuator based on this property is demonstrated. Nanotubes paper consists of bundles of nanotubes having their long axis lying in the plane of the paper, but randomly oriented in the plane. The actuator consisted of 3 × 20 mm strips of nanopaper 25–50 μm thick. The two strips are bonded to each other by double stick Scotch type. An insulating plastic clamp at the upper end supports, the paper and holds the electrical contacts in place. The sheets were placed in a one molar NaCl solution. Application of a few volts produced a deflection of upto a cm, and could be reversed by changing the polarity of the voltage. Application of an AC voltage produced an oscillation of the cantilever. This kind of actuator is called a bimorph cantilever actuator because the device response depends on the expansion of opposite electrodes. This actuator is not a NEMS or MEMS. But it works on the effect of charging on the individual carbon nanotubes. One can assemble 3 fibers with their axes parallel and in contact. The outer tubes would be metallic and the inner tube insulating.

Electron beam lithography is used to fabricate silicon structures <10 nm size. There is a problem of communicating with and sensing the motion of the nanoscale devices. Another problem is our understanding of mechanical behaviour of objects which have 10% of their atoms on or near the surface. Consider a beam of length 10nm and thickness of 1nm will have a resonant frequency 10^5 times greater of the order of 20–30GHz. As the frequency increases the displacements of the beam can range from a picometer (10^{-12}M) to 10^{-15}M. these high frequencies and small displacement are very difficult to detect.

Optical reflection methods such as those used in the μM range on the cantilever tips of STM are not applicable because of the diffraction limit. This occurs when the size of the object from which light is reflected becomes smaller than the wavelength of the light. Transducers are generally used in MEMS devices to detect motion. The MEMS accelerometer is an example of the detection of motion using a transducer. In the accelerometer mechanical motion is detected by a change in capacitance which can be measured by an electrical circuit. The limitations in the development of NEMS is to develop transducers to sense displacement as small as 10^{-15} to 10^{-12}M. and do so at frequencies upto 30GHz.

However the small effective mass of a nm sized beam renders its resonant frequency extremely sensitive to slight changes in its mass. This frequency can be affected by adsorption of a small number of atoms on the surface, which forms a basis for a variety of very high sensitive sensors.

Nanosized cantilevers have very high Q (quality factor) for the resonance. $Q = W_0 / \Delta W$ where ΔW is the width of the resonance at half very height and W_0 is the resonant frequency. The quality factor is the energy stored divided by the energy dissipated per cycle so the inverse of the quality factor 1/Q is a measure of the dissipation energy. High Q devises also have low thermo mechanical fluctuations. The Q values of high Q electrical devices are in the order of several hundreds but NEMS oscillations can have Q values 1000 times higher. Another advantage of NEMS devices is that they require very little power to drive them. A 10^{-12}W of power can drive an NEMS device with a low signal to noise ratio (SNR).

Nanogears are fabricated using computation. The "teeth" of the gear would be C_6H_6 molecules bonded to the outer walls of the tube. The power gear is charged in order to make a dipole moment across the diameter of the tube. Application of an alternating field could induce this gear to rotate. Any nanosized machine should be capable of moving the gear.The idea of using an electric field which does not require contacts with the nanostructure to roll a C_{60} molecule over a flat surface is known.

Figure 4.4 Computer simulation has been used to evaluate the potential of various nanomachine concepts. One example, shown in Figure 4.4 is the idea of making gears out of nanotubes. The "Teeth" of the gear would be C_6H_6 molecules bonded to the outer walls of the tube. The power gear on the left side of the figure is charged in order to make a dipole moment across the diameter of the tube. Application of an alternating electric field could induce this gear to rotate. An essential part of any nanosized machine is the capability of moving the power gear. The idea of using an electric field which does not require contacts with the nanostructure to roll a C_{60} molecule over a flat surface is proposed.

An isolated C_{60} molecule is adsorbed on the surface of an ideally flat ionic crystal such as KCl. The application of an electric field would polarize the C_{60} molecular putting +ve and –ve charges on opposite sides of the sphere. Because the molecule has a large polarizability and a large diameter, a large electric dipole moment is induced. If the interaction between the dipole moment and the applied electric

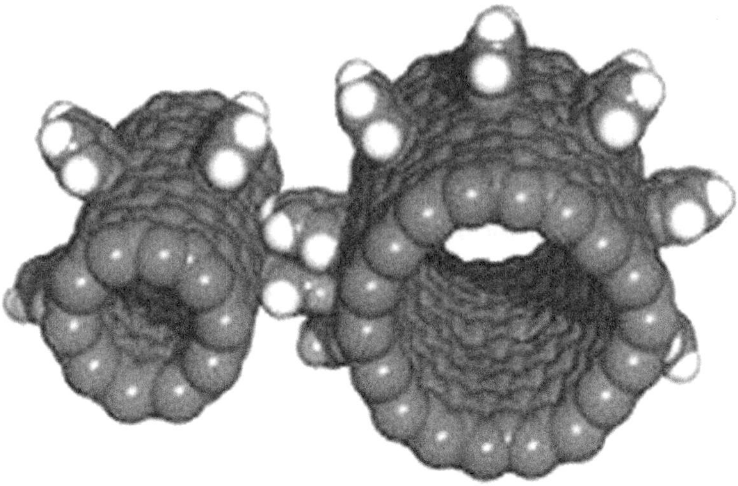

Fig. 4.4 Nanogears – A model-bezene molecule attached to the outside carbon nanotubes.

field is greater than the interaction between the moment and the surface of the material, rotation of the electric field should casue the C_{60} molecule to roll across the surface.

AFM employs a sharp tip mounted on a cantilever spring which is scanned closely over the surface of a material. The deflection of the cantilever is measured. In the region of surface atoms the deflection is larger because of the layer interaction between the tip and the atoms. The cantilever are fabricated by photolithographic methods from Silicon, SiO_2 or SiN. They are typically 100μm long 1μm thick and have spring constants between 0.1 to 1.0 N/M (Newton/meter). Operating in the tapping mode where the change in the amplitude of an oscillating cantilever driven near its resonance frequency is measured by as the tip taps the surface, can increase the sensitivity of the instruments.

A MWNT may be bonded to the side of a tip of a conventional silicon cantilever using a soft acrylic adhesive. If the nanotube crashes into the surface generating a force greater than the Euler buckling force, the nanotube does not break but rather bends away and then snaps back to its original position. The nanotubes tendency to buckle rather than break makes it unlikely that the tip will break. The MWNT tip can also be used in the tapping mode. When the nanotubes bends on impact, there is a coherent de excitation of the Cantilever oscillation. The MWNT serves as a complaint spring which moderates the impact of each top on the surface. Because of the small cross-section of the tip, it can reach into deep trenches on the surface that are in accessible to normal tips. Since MWNT's are electrically conducting, they may also used as probes for an STM.

An artificial single molecule machine that converts light energy to physical work. Azobenzene can change from the trans isomer to the cis isomer by subjecting it to 313nm light. A chain of azobenzene molecules form a polymer. By exposing to 365nm light this polymer can go from trans to cis. The molecular machines attached to the trans form of the polymer to the cantilever of an AFM. They are subjected to 365nm wavelength causing the polymer to contract and the beam to bend. Exposure to 420nm light causes the polymer to the trans form allowing the beam to return to its original position. By alternatively exposing the polymer to pulses of 425 & 365 nm light, the beam can be made to oscillate.

4.6 NANOSWITCHES

Nanosize architecture is becoming difficult and expensive. This has motivated an effort to synthesis molecules which display switching behaviour. This behaviour form the basis for information storage and logic circuiting in computers using binary systems. A molecule A that can exist in two different states such as two different conformations A and B and can be converted to reversibly between the two states by external stimuli such as light or a voltage can be used to store information. In order for the molecule to be used as a zero or one digit state, necessary binary logic, the change between the states must be fast and reversible by external stimuli. The two states must be thermally stable and be able to switch back and forth many times. The states must be distinguishable by some probe and the application of the probe R is called the "Read mode". Electrochemical oxidation and reduction increases the thermal stability of azo benzene. The cis isomer is reduced by the addition of H atoms at a more anodic (–ve) potential and then converted back to the trans isomer by oxidation which removes the H atoms.

A chirooptical molecular switch uses circularly polarized light [CPL] to bring about changes between isomers. The application of left circularly polarized light (–) CPL to the molecular conformation M causes a rotation of the 4 ring group from a right handed helical structure to a left handed helical arrangement P. Right circularly polarized light (+) CPL brings about the reverse transformation. Linearly polarized light (LPL) can be used to read the switch by monitoring the change in the axis of the light polarizer. The system can be erased using unpolarized light (UPL).

Molecules undergo conformational changes involving rearrangements of the bonding in a molecule can also be the basis of molecular switching. When the colourless spiropyran is subjected to UV light, the C–O bond opens forming merocyanine. When the

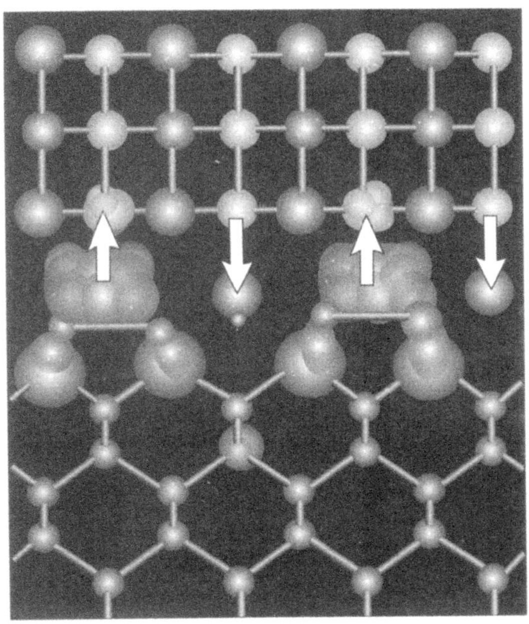

Fig. 4.5 Nanoswitches

merocyanine is subjected to visible (red) light, hv_2 or heat (Δ) the spiropyran reforms. Nanosize architecture is becoming difficult and expensive. This has motivated an effort to synthesis molecules which display switching behaviour. This behaviour form the basis for information storage and logic circuiting in computers using binary systems. One such nanoswitch is shown in **Figure 4.5.**

A catenane molecule is used to make a molecular switch that can be turned on and off with the application of voltage. A catenane is a molecule with a molecular ring mechanically interlinked with another molecular ring. It has two different Switched states. This molecule is 0.5nm long and 1nm wide. A monolayer of the catenane anchored with amphiphilic phospholipids counterions is sandwitched between the two electrodes. The structure is the open switch position as the configuration does not conduct electricity. When the molecule is oxidized by applying a voltage which removes an electron, the tetrathia fulvalene group which contains the sulphur, becomes positively ionized and is thus electro statically repelled by the cyclophane group, the ring containing the N_2 atoms. This causes the change in structure, which essentially involves a rotation of the ring of the left side molecule to the right side.

The STM is used to measure the conductivity of long chain like molecules. A monolayer of octanethiol was formed on a gold surface by self assembly. The S group at the end of the molecule bonded to the surface. Some of the molecules were removed by a solvent technique and replaced with 1,8 octanedithiol, which has S groups at both ends of the chain. A gold coated STM tip was scanned over the top of the monolayer to find the 1,8 octanedithiol. The tip was then put in contact with the end of the molecule forming an electric circuit between the tip and the flat gold surface. The octanethiol molecules which are bound only to the bottom of the gold electrode serve as molecular insulators, electrically isolating the octanedithiol wires. The voltage between the tip and the bottom gold electrode is then increased and the current measured. One observes 5 distinct families of curves, each an integral multiple of the fundamental curve the fundamental curve corresponds to electrical conduction through a single dithiol molecule, the other curves correspond to conduction through two or more such molecules. It should be noted that the current is Quite low (nA) and the resistance of the molecule is estimated to be 900mΩ.

In order to use a molecular as an on/off switch, a molecule which contains a thiol group (–SH) that can be attached to gold by losing a H atom. The molecule 2-amino–4-ethylnyl phenyl–5-nitro–1-benzene thiolate consists of 3 benzene rings linked in a row has an amino (NH_2) group which is an electron donor, pushing electric charge towards the ring. On the other side is an electron acceptor nitro (NO_2) group which withdraws electrons from the ring. The net result is that centre ring has a large electric dipole moment.

An electronic switch may be made using a conducting molecule bonded at each to the gold electrodes. There is an onset of current at 1.6V then a pronounced increase followed by a sudden drop at 2.1V. The result was observed at 60K but not at room temperature. The effect is called "Negative-differential resistance". The proposed mechanism for the effect is that the molecule is initially nonconductive and at the voltage where a current peak is observed the molecule gains an electron, forming a radical ion and becomes conducting. As the voltage is increased further a second electron is added and the molecule forms a non-conducting dianion.

4.7 NANOCOMPUTERS

The molecular switches have to be connected together to form logic gates. A roxatane molecule which changes conformation when it gains and loses electron by rotation of the O_2 ring. The assembly consists of a monolayer of rotaxane molecules sand witched between two parallel upper electrodes has a large of titanium (Ti) on it and the lower one has an alumina (Al_2O_3) layer that acts as a tunneling barrier. To fabricate this switch, an aluminium electrode was made by lithographically pattering 0.6μM diameter aluminium wires on a silica substrates. The electrode was then exposed to oxygen to allow a layer of Al_2O_3 to form on it. Next a single monolayer of the roxatane molecule was deposited as a Langmuir Blodgett film. Then a 5μM layer of titatium followed by a thicker layer of aluminum wire were evaporated through a contact mask using electron beam deposition.

The current-voltage characteristics is explained as follows : An application of –2V the read mode, caused a sharp increase in current. The switch could be opened by applying 0.7V. The difference in the current between the open and closed switch was a factor of 60–80. A number of these switches were wired together in arrays to form logic gates. Two switches (A & B) connected can act as an AND gate. In an AND gate both switches have to be on for an output voltage to exist. There should be little or no response when both switches are off or only one switch is on. Molecular switching devices form a base of future computer technology. **Figure 4.6** presents nanocomputing taxonomy spectrum.

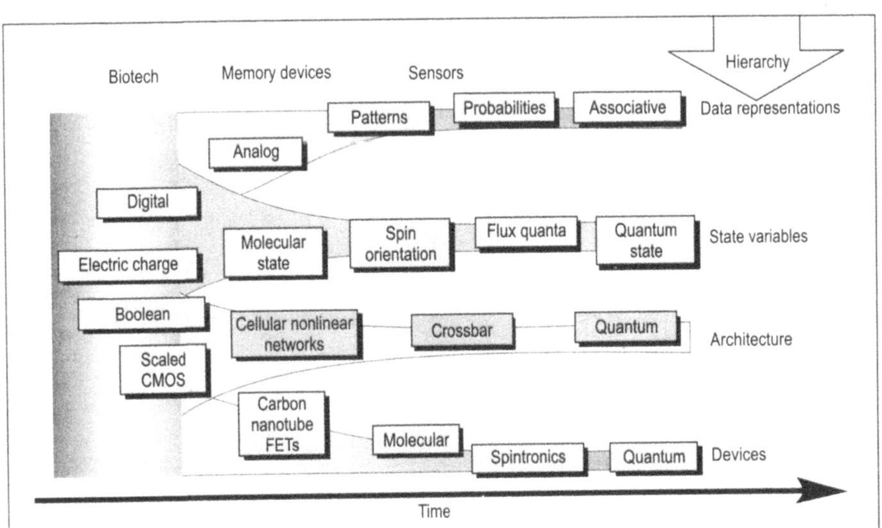

Fig. 4.6 Nanocomputing taxonomy.

4.8 NANOFILTERS

Nanofiltration is a technique that has prospered over the past few years. It is applied mainly in drinking water purification. Process steps such as water softening, decolouring and micropollutant control. Viruses of the dimensions 10–100 nm are removed by ultrafiltration. It is an low pressure membrane process that separates materials in the 0.001 to 0.1 μm range. Dissolved salts and colour producing organic compounds are removed by revrese osmosis and Nanofiltration.

Nanoceramic filters are used for water purification. The filters are a mixture of nanoalumina fiber and microglass. Nanoceramic filters have asymmetrically shaped pores rather than cylindrical pores. Air flow and bubble pressure measurements show that the average pore size is between 1 and 2 μm. Nanoceramic fibers are highly positively charged and can retain negatively charged particles. Usually Nano ceramic filter has a removal efficiency of 99.99% for virus and bacteria. They have several orders of magnitude flow capability compared to equivalent membrane filters. They have higher capacity for particulates and less clogging. They chemisorb dissolved heavy metals. They are very porous (~ 85%). The filter is very difficult to clog or foul due to its wide open structure. They filter RNA and DNAs. Nanofiltration systems are applied to remove heavy metals, salt contents, particles, heavy metals from waste waters.

❑❑❑

Bibliography

Chapter 1

1. P.M. Ajayan. "Carbon Nanotubes", Handbook of Nano Structured Materials and Nanotechnology, H.S. Nalwa ed., Academic Press, San Dugo, 2000, Vol.5, Ch 6, P.375.
2. P.G. Collins and P. Avouris, "Carbon Nanotubes", Sci. Am 62 (Dec. 2000).
3. M.S. Dresselhaus, G. Dresselhaus and R. Saito, "Nanotechnology in Carbon Materials", in Nanotechnolgoy G. Timp Ed., springer–Verlag, 1006, Ch 7, P.285.
4. T.W. Ebbesen, "Carbon Nanotubes", Phys. Today, 26 (June 1996).
5. R. Saito, G. Dresselhaus and M.S. Dresselhaus, "Physical Properties of Carbon Nanotubes", Imperical College Press, London 1999.
6. M.S. Feld and K. An "Single Atom Laser", Sci. Am 57 (July 1998).
7. M. Henini, "Quantum Dot Nanostructures", Materials Today, 48 (June 2002).
8. L. Jacak, P. Hawrylak and A. Wejs, "Quantum Dots", Springer, Berlin, 1998.
9. A. Archut and F. Vogtle, "Dendritic Molecules—Historic Development and Future Applications", in Nalwa (2000) Vol. 5, Ch 5, p.333.
10. H. Kasai, H.S. Nalwa, S. Okada, H. Oikawa and H. Nakarishi, "Fabrication and Spectroscopic Characterization of Organic Nanocrystals", in Nalwa (2000) Vol 5, Ch 8, p.334.

Chapter 2

1. R.P. Anders et al, "Research Opportunities in Clusters and Cluster Assembled Materials", J. Mater. Res. 4, (1989) 704.
2. W.A. De Heer, "Physics of Simple Metal Clusters", Rev. Mod. Phys. 65, 1993, 611.

3. M.A. Duncan and D. H. Rouray, "Microclusters", Sci. Am. 110 (Dec 1989).
4. S.N. Khanna, "Handbook of Nanophase Materials", in A.N. Goldstein Ed. Marcel Dekkar, NY, 1997, Ch. 1.
5. M. Morse, "Clusters of Transition Atoms", Chem. Rev. 86 (1986) 1049.
6. S. Sugano and H. Koizumi, "Microcluster Physics" , Springer-Verlag, Heidelberg, 1998.

Chapter 3

1. J.M. Cowley and J.C.H. Spence, "Nano Diffraction" in Nalwa (2000) Vol 2, Ch. 1, 2000.
2. H.S. Nalwa, Ed., "Handbook of Nano Structured Materials and Nanotechnology, Vol.1, 405, Academic Press, Boston 2000.
3. R.E. Whan, "Materials Characterization", Vol.10, of Metals Handbook, American Society of Metals, Metals Park OH. 1986.
4. H.S. Nalwa Ed. "Handbook of Nano Structured Materials and Nanotechnology," Vol 2, Spectroscopy and Theory Vol 4, Optical Properties, Academic Press, Isan Diego, 2000.
5. D. Ruger and P. Hansma, "Atomic Force Microscopy", Phys, Today 43, (Oct 1990) 23.
6. J.A. Stroscio and D. M. Eigler, "Atomic and Molecular Manipulation with a Scanning Tunneling Microscope", Science, 254 (1991) 1319.

Chapter 4

1. W.R. Moser ed., "Advanced Catalysts and Nano Structured Materials", Academic Press, Sandiego, 1996.
2. M. Gross, "Travels to the Nanoworld", Plenum, Ny 1999.
3. H.S. Nalva Ed., "Handbook of Nano Structured Materials and Nanotechnology" Vol 5, Organics. Polymers and Bedogical Compounds, Academic Press, Boston 2000.
4. D. Bishop, P. Gammel and C.R. Giles, "Little Machines that are Making it Big", Phys. Today 54 (Oct 2001) 38.
5. R. Dagani, "Building from the Bottom Up", Chem. Engg. News, 28 (Oct. 2000).
6. M. A. Red and J. M. Tour, "Computing with Molecules", Sci. Am. 38, (June 2000) 86.
7. M. Roukes, "Plenty of Room Indeed" Sci. Am, 39, (Sept. 2001) 48.
8. M. Roukes, "Nanoelectromechanical Systems Face the Future", Phys. World. (Feb 2001).
9. G.M. White side and J.C. Love, "The Art of Building Small", Sci. Am 39 (Sep 2001) 285.

Index

❑❑❑